WORKSHEETS FOR CLASSROOM OR LAB PRACTICE

CHRISTINE VERITY

MATHEMATICS IN ACTION: PREALGEBRA PROBLEM SOLVING

THIRD EDITION

The Consortium for
Foundation Mathematics

Addison-Wesley
is an imprint of

Reproduced by Pearson Addison-Wesley from electronic files supplied by the author.

Publishing as Addison-Wesley, 75 Arlington Street, Boston, MA 02116.

ISBN-13: 978-0-321-73837-0
ISBN-10: 0-321-73837-3

1 2 3 4 5 6 OPM 15 14 13 12 11

Addison-Wesley
is an imprint of

www.pearsonhighered.com

Table of Contents

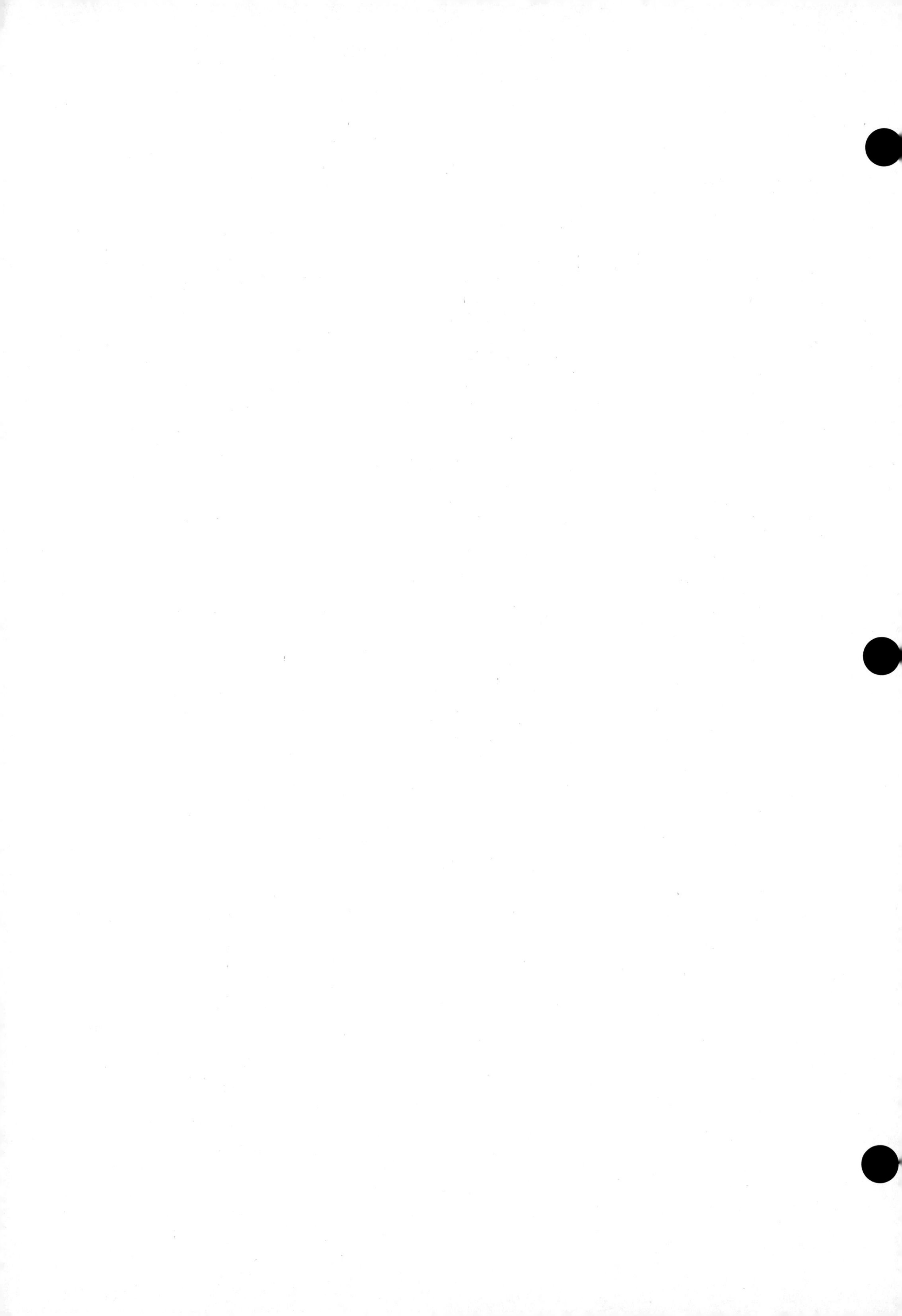

Name: Date:
Instructor: Section:

Chapter 1 WHOLE NUMBERS

Activity 1.1

Learning Objectives

1. Read and write whole numbers.
2. Compare whole numbers using inequality symbols.
3. Round whole numbers to specified place values.
4. Use rounding for estimation.
5. Classify whole numbers as even or odd, prime, or composite.
6. Solve problems involving whole numbers.

Key Terms

Use the vocabulary terms listed below to complete each statement in Exercises 1–7.

even	**odd**	**place value**	**factors**
prime	**composite**	**rounding**	

1. ____________________ a given number to a specified place value is used to approximate its value.

2. A ____________________ number is a whole number greater than 1 and divisible only by itself and 1.

3. ____________________ numbers are whole numbers that are exactly divisible by 2.

4. ____________________ numbers are whole numbers that are not exactly divisible by 2.

5. The ____________________ of a number are all the numbers that divide evenly into the given number.

6. ____________________ numbers are whole numbers greater than 1 that have other factors as well as 1 and itself.

7. Each digit in a numeral has a ____________________ determined by its relative placement in the numeral.

Practice Exercises

For #8-11, write each number as a numeral.

8. thirty-nine thousand five hundred eighteen

9. six million, two hundred ninety-one thousand eight hundred thirty four

8. ________________

9. ________________

10. twenty-six thousand nine hundred seven

11. eight hundred fifty-two thousand twenty-three

10. ________________

11. ________________

For #12-15, round each number as indicated.

12. Round 96,542 to the nearest thousand.

13. Round 563,297 to the nearest hundred.

12. ________________

13. ________________

14. Round 8,232,017 to the nearest ten-thousand.

15. Round 9,643,278 to the nearest million.

14. ________________

15. ________________

For #16-19, use >, <, or = to compare the numbers.

16. 670,934 and 671,213

17. 5,838,721 and 5,829,466

16. ________________

17. ________________

18. 999,010 and 1,090,090

19. 169,329 and 169,511

18. ________________

19. ________________

Name: Date:
Instructor: Section:

For #20-21, write the number in words.

20. 56,210

20. ____________

____

21. 207,934

21. ____________

____

For #22-24, determine whether each number is even or odd, and give a reason.

22. 96,737

23. 23,663

22. ____________

23. ____________

24. 1,808,456

24. ____________

For #25-28, determine whether each number is prime or composite, and give a reason.

25. 95

26. 37

25. ____________

26. ____________

27. 23

28. 57

27. ____________

28. ____________

Concept Connections

29. Examples of consecutive even whole numbers are 0, 2, 4, 6, 8, and so on. What is the common difference between an even whole number and the next consecutive even whole number?
Examples of consecutive odd whole numbers are 1, 3, 5, 7, 9, and so on. What is the common difference between an odd whole number and the next consecutive odd whole number?

____

____

30. After writing a check to pay off a credit card, your credit card company noted that the amount listed as a number did not match the amount listed as a numeral. You had written the number as $472.65 and written the numeral as four hundred seventy dollars and sixty-five cents. What was your error?

____

____

Name: Date:
Instructor: Section:

Chapter 1 WHOLE NUMBERS

Activity 1.2

Learning Objectives
1. Read tables.
2. Read bar graphs.
3. Interpret bar graphs.
4. Construct graphs.

Key Terms

Use the vocabulary terms listed below to complete each statement in Exercises 1–3.

graphing grid **horizontal** **vertical** **input** **output**

1. The second number in an ordered pair is always found or given along the ________________ direction on a graph and is called a(n) ________________ value.

2. A(n) ________________ displays paired data (x-coordinate, y-coordinate).

3. The first number in an ordered pair is always found or given along the ________________ direction on a graph and is called a(n) ________________ value.

Practice Exercises

For #4-7, use the following grid.

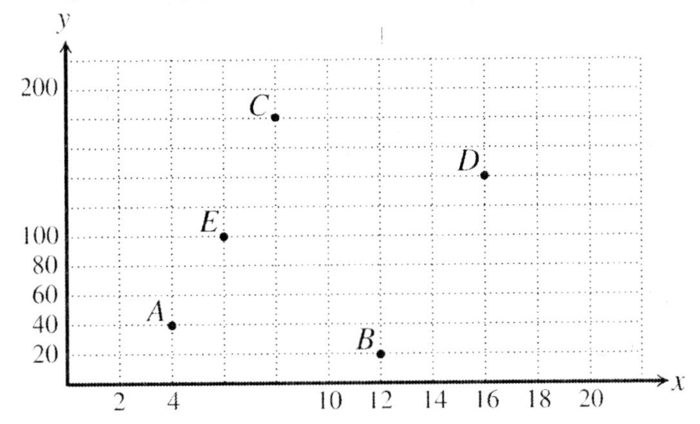

4. What is the size of the input units on the grid shown?

5. What is the size of the missing output units on the grid shown?

6. What are the missing input units?

7. What are the missing output units?

4. ________________

5. ________________

6. ________________

7. ________________

For #8-19, use the following grid.

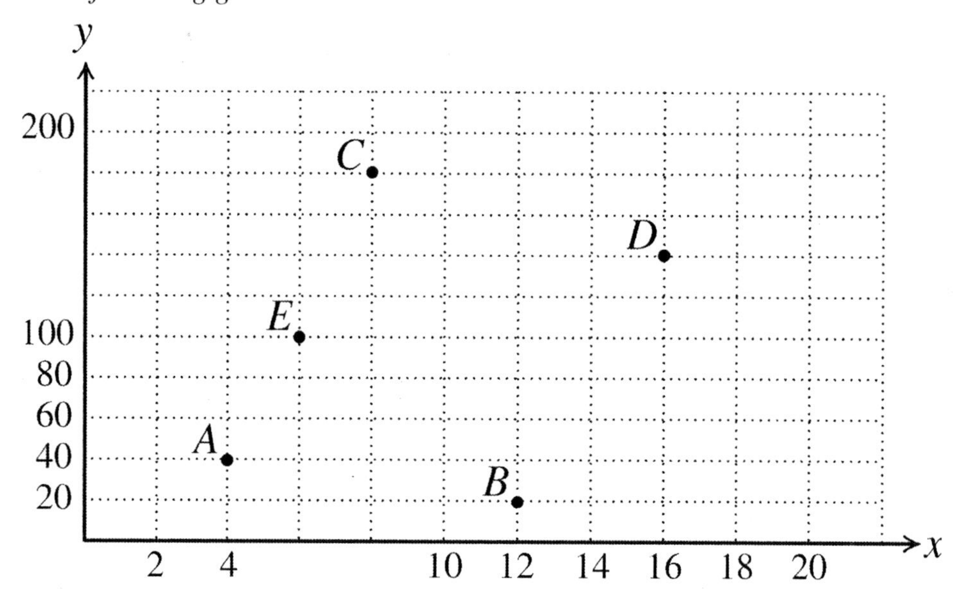

8. What are the coordinates of point *A* given on the grid?

9. What are the coordinates of point *B* given on the grid?

10. What are the coordinates of point *C* given on the grid?

11. What is the x-coordinate of the point *D*?

12. What is the y-coordinate of the point *E*?

13. Write in the missing input and output units on the graph.

8. ________________

9. ________________

10. ________________

11. ________________

12. ________________

For #14-19, plot each point on the grid and write the ordered pair next to the point.

14. (10, 80)

15. (18, 60)

16. x-coordinate is 14 and y-coordinate is 200

17. x-coordinate is 2 and y-coordinate is 160

18. (0, 0)

19. (20, 0)

Name: Date:
Instructor: Section:

For #20-23, graph the data in the table as a bar chart on a grid.

U.S. Cities	Average Daily Temperature, Dec 2010
Miami	62
Atlanta	38
Washington, D.C.	34
Chicago	21

20. Write the name of the output along the vertical axis and the name of the input along the horizontal axis.

21. List the input categories along the horizontal axis.

22. List the units along the vertical axis.

23. Draw the bars corresponding to the given input/output pairs.

For #24-28, the following chart shows data from a survey of New Year's resolutions.

Weight loss
Eating healthier
Being more active
Quit smoking

5 10 20 30 40 50 60 70
Survey Percent

24. Which resolution has the highest percent? Estimate the number.

25. Which resolution has the lowest percent? Estimate the number.

24. ______________

25. ______________

26. Which resolution is between 40% and 50%? Estimate the number.

27. Which resolution is between 50% and 60%? Estimate the number.

26. ______________

27. ______________

28. What feature in this chart aids in estimating the percentage of the resolution?

28. ______________

____

Concept Connections

29. What is the significance of the order in an ordered pair?

____

____

30. What is the difference between (2, 3) and (3, 2)?

____

____

Name: Date:
Instructor: Section:

Chapter 1 WHOLE NUMBERS

Activity 1.3

Learning Objectives

1. Add whole numbers by hand and mentally.
2. Subtract whole numbers by hand and mentally.
3. Estimate sums and differences using rounding.
4. Recognize the associative property and the commutative property for addition.
5. Translate a written statement into an arithmetic expression.

Key Terms

Use the vocabulary terms listed below to complete each statement in Exercises 1–8.

addends	**sum**	**commutative**	**associative**
estimation	**subtrahend**	**difference**	**minuend**

1. The number that is subtracted is called the ________________ .

2. The number being subtracted from is called the ________________ .

3. Numbers that are added together are called ________________ .

4. The result of a subtraction is called the ________________ .

5. The ________________ property of addition is demonstrated in the calculation of a sum of three numbers, it does not matter if the first two numbers or the last two numbers are added together first.

6. The total of numbers is called the ________________ .

7. The ________________ property of addition is demonstrated when the order in which two numbers are added does not matter.

8. ________________ is useful for adding or subtracting numbers quickly and to check the reasonableness of an exact calculation.

Practice Exercises

For #9-10, determine the sum using Method 1.

9.
$$\begin{array}{r} 729 \\ +\ 234 \\ \hline \end{array}$$

10.
$$\begin{array}{r} 32 \\ 564 \\ +\ 87 \\ \hline \end{array}$$

9. ________________

10. ________________

For #11-12, determine the sum using Method 2.

11. $\begin{array}{r} 832 \\ +\ 149 \\ \hline \end{array}$

12. $\begin{array}{r} 62 \\ 394 \\ +\ 18 \\ \hline \end{array}$

11. ______________

12. ______________

13. Determine the sum of 58 and 64.

14. Determine the sum of 64 and 58.

13. ______________

14. ______________

15. Are the sums the same for exercises #13 and #14?

16. What property of addition is demonstrated by exercises #13 and #14?

15. ______________

16. ______________

For #17-18, determine the sums. Do the addition in the parentheses first.

17. $37+(18+61)$

18. $(37+18)+61$

17. ______________

18. ______________

19. Are the sums the same for exercises #17 and #18?

20. What property of addition is demonstrated by exercises #17 and #18?

19. ______________

20. ______________

Name: Date:
Instructor: Section:

For #21-22, use the sum: 183 + 95 + 314.

21. Estimate the sum.

22. Determine the actual sum.

21. ________________

22. ________________

For #23-24, find the difference.

23. $462 - 136$

24. $785 - 273$

23. ________________

24. ________________

For #25-28, translate each into an arithmetic expression.

25. 16 plus 37

26. 12 subtracted from 36

25. ________________

26. ________________

27. 63 less than 250

28. 37 decreased by 19

27. ________________

28. ________________

Concept Connections

29. From Exercise #21, was the estimate higher, lower, or the same as the actual sum (from Exercise #22)? Is it possible to determine whether an estimate will be higher, lower, or the same as the actual sum before calculating the actual sum? How?

____

____

30. Generally, is subtraction commutative? Give one example when subtraction is not commutative and one example, a special case, of when subtraction is commutative.

____

____

Name: Date:
Instructor: Section:

Chapter 1 WHOLE NUMBERS

Activity 1.4

Learning Objectives

1. Multiply whole numbers and check calculations using a calculator.
2. Multiply whole numbers using the distributive property.
3. Estimate the product of whole numbers by rounding.
4. Recognize the associative and commutative properties for multiplication.

Key Terms

Use the vocabulary terms listed below to complete each statement in Exercises 1–4.

distributive	**associative**	**commutative**
factor	**product**	

1. In the equation $5 \times 7 = 35$, the 35 represents the ________________ and 5 and 7 represent the ________________ (s).

2. One example of the ________________ property is $2 \cdot 9 = 9 \cdot 2$.

3. One example of the ________________ property is $4 \cdot (6+3) = 4 \cdot 6 + 4 \cdot 3$.

4. One example of the ________________ property is $12 \cdot (8 \cdot 5) = (12 \cdot 8) \cdot 5$.

Practice Exercises

For #5-12, multiply vertically.

5. $\begin{array}{r} 36 \\ \times\ 4 \\ \hline \end{array}$

6. $\begin{array}{r} 537 \\ \times\ 7 \\ \hline \end{array}$

5. ________________

6. ________________

7. $\begin{array}{r} 89 \\ \times\ 6 \\ \hline \end{array}$

8. $\begin{array}{r} 906 \\ \times\ 8 \\ \hline \end{array}$

7. ________________

8. ________________

9. $\begin{array}{r} 65 \\ \times\ 43 \\ \hline \end{array}$

10. $\begin{array}{r} 3021 \\ \times\ \ 35 \\ \hline \end{array}$

9. ______________

10. ______________

11. $\begin{array}{r} 85 \\ \times\ 32 \\ \hline \end{array}$

12. $\begin{array}{r} 1896 \\ \times\ 204 \\ \hline \end{array}$

11. ______________

12. ______________

13. Multiply 9 and 38 by rewriting 38 as 30 + 8 and use the distributive property to obtain the result.

13. ______________

14. Multiply 9 and 38 vertically.

14. ______________

15. Multiply 14 and 48 by rewriting 48 as 50 – 2 and use the distributive property to obtain the result.

15. ______________

16. Multiply 14 and 48 vertically.

16. ______________

For #17-18, four 500-count packages of napkins are purchased.

17. Use addition to determine the total number of napkins.

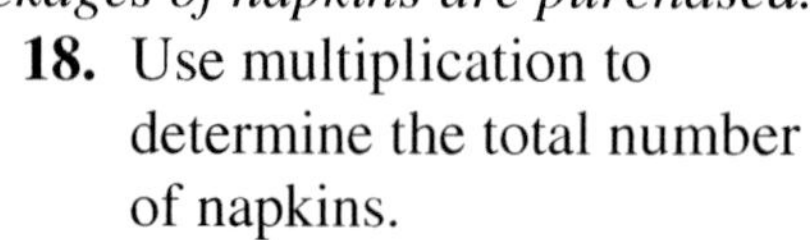

18. Use multiplication to determine the total number of napkins.

17. ______________

18. ______________

Name: Date:
Instructor: Section:

For #19-21, ten containers of 50-count toothpicks are purchased.

19. Determine the total number of toothpicks by calculating $10 \cdot 50$.

20. Determine the total number of toothpicks by calculating $50 \cdot 10$.

19. ______________

20. ______________

21. What property of multiplication is demonstrated by the fact that the answers in #19 and #20 are the same?

21. ______________

For #22-23, evaluate by finding the product in parentheses first.

22. $8 \cdot (17 \cdot 30)$

23. $(8 \cdot 17) \cdot 30$

22. ______________

23. ______________

24. What property do the results in #22 and #23 demonstrate?

24. ______________

For #25-26, evaluate.

25. $63 \cdot 42$

26. $42 \cdot 63$

25. ______________

26. ______________

27. Are the answers to #25 and #26 the same?

28. What property do the results of #25 and #26 demonstrate?

27. ______________

28. ______________

Concept Connections

29. Three kids are given equal piles of pennies and are told to count them. The first child, in first grade, proceeds to count each one until he reaches 90. The second child, in fourth grade, creates rows of 10 pennies, and declares that ten times nine OR nine times ten is 90. The third child, in eight grade, creates five stacks of ten pennies, then four stacks of ten pennies, and declares that five times ten is 50, four times ten is 40, and therefore 50 plus 40 is 90, OR, four plus five is nine, and nine times ten is 90. What mathematical operations or properties are demonstrated by each child?

____

____

30. How is factoring related to the distributive law property of multiplication?

____

____

Name: Date:
Instructor: Section:

Chapter 1 WHOLE NUMBERS

Activity 1.5

Learning Objectives
1. Divide whole numbers by grouping.
2. Divide whole numbers by hand and by calculator.
3. Estimate the quotient of whole numbers by rounding.
4. Recognize that division is not commutative.

Key Terms
Use the vocabulary terms listed below to complete each statement in Exercises 1–5.

undefined	**quotient**	**remainder**	**dividend**
divisor	**zero**	**one**	**inverse**

1. In the expression $43 \div 6$, we get 7 with 1 leftover. To check our work, we do the calculation $(7 \times 6) + 1 = 43$, where 6 represents the ____________________ , 7 represents the ____________________ , 1 represents the ____________________ , and 43 represents the ____________________ .

2. Multiplication and division are ____________________ operations.

3. Zero divided by any nonzero whole number is ____________________ .

4. Any nonzero whole number divided by zero is ____________________ .

5. Any nonzero whole number divided by itself is ____________________ .

Practice Exercises
For #6-15, calculate and identify the quotient and remainder (if any).

6. $63 \div 9$

7. $117 \div 3$

6. ____________________

7. ____________________

8. $98 \div 4$

9. $243 \div 12$

8. ____________________

9. ____________________

10. $516 \div 1$

11. $697 \div 27$

10. ________________

11. ________________

12. $0 \div 23$

13. $19 \div 0$

12. ________________

13. ________________

14. $904 \div 904$

15. $699 \div 21$

14. ________________

15. ________________

For #16-19, while camping, a family of five share equally six 8-pack boxes of instant oatmeal.

16. Determine the total number of packets to be shared.

17. How many packets will each family member receive?

16. ________________

17. ________________

18. Set the calculations up as a long division and divide.

19. Do this problem again using repeated subtraction.

18. ________________

19. ________________

Name: Date:
Instructor: Section:

20. Divide $28 \div 4$.

21. Do you get the same result as #20, by doing the division $4 \div 28$?

20. ______________

21. ______________

22. Does the commutative property hold for division? That is, does $28 \div 4 = 4 \div 28$?

22. ______________

For #23-25, evaluate $4937 \div 42$.

23. Estimate $4937 \div 42$

24. Find the exact answer for $4937 \div 42$.

23. ______________

24. ______________

25. Is your estimate higher, lower, or the same, as compared to the actual answer?

25. ______________

26. Estimate $32,900 \div 593$.

27. Find the exact answer for $32,900 \div 593$.

26. ______________

27. ______________

28. Is your estimate higher, lower, or the same, as compared to the actual answer?

28. ______________

Concept Connections

29. You are responsible for bringing hot dogs and rolls to a party. You expect 50 people to each eat one hotdog and roll. There are ten hotdogs in each package and eight rolls in each package. How many packages of hotdogs and rolls do you need to purchase? What will be leftover after the party?

____

____

30. An intern for a TV game show is estimating how many trips a contestant must make to move 10,000 tennis balls, carrying seven tennis balls at a time. To overestimate the intern uses five tennis balls, and to underestimate the intern uses ten tennis balls. Which estimate is closer to the actual number of trips?

____

____

Name: Date:
Instructor: Section:

Chapter 1 WHOLE NUMBERS

Activity 1.6

Learning Objectives
1. Use exponential notation.
2. Factor whole numbers.
3. Determine the prime factorization of a whole number.
4. Recognize square numbers and roots of square numbers.
5. Recognize cubed numbers.
6. Apply the multiplication rule for numbers in exponential form with the same base.

Key Terms

Use the vocabulary terms listed below to complete each statement in Exercises 1–8.

square root	**perfect square**	**square**	**perfect cube**
cube	**base**	**exponent**	**power**
zero	**exponential form**	**one**	

1. A ____________________ is a rectangle in which all the sides have equal length.

2. A ____________________ is a box for which all of the edges have equal length.

3. A whole-number ____________________ indicates the number of times to use the ____________________ as a factor. A number written as 10^{14} is in ____________________. The expression 10^{14} is called a ____________________ of 10.

4. A whole number is a ____________________ if it can be written as the product of three whole-number factors that are equal.

5. A whole number is a ____________________ if it can be rewritten as the product of two whole-number factors that are equal.

6. The ____________________ of a whole number is one of the two equal factors whose product is the whole number.

7. A prime number is a whole number greater than 1 whose only factors are itself and ____________________.

8. Any nonzero whole number raised to the ____________________ power equals 1.

Concept Connections

29. State the Fundamental property of whole numbers.

---

30. Explain how to multiply numbers written in exponential form that have the same base.

Name: Date:
Instructor: Section:

Chapter 1 WHOLE NUMBERS

Activity 1.7

Learning Objectives
1. Use order of operations to evaluate arithmetic expressions.

Practice Exercises

For #1-4, evaluate each expression.

1. $6(30+4)$

2. $6 \cdot 30 + 4$

1. ____________

2. ____________

3. $6 \cdot 30 + 6 \cdot 4$

4. $30 + 6 \cdot 4$

3. ____________

4. ____________

5. For questions #1-4, which two of the preceding arithmetic expressions have the same answer?

5. ____________

6. State the property that produces the same answer for the pair of expressions.

6. ____________

For #7-26, perform the calculations without a calculator. After solving, use a calculator to check your answers.

7. $51 \div 3 + 7$

8. $56 \div 7 - 2 \cdot 4$

7. ____________

8. ____________

9. $17+40\div 8\cdot 3-11$

10. $28+5\cdot 4-18\div 6$

9. ______________

10. ______________

11. $54/(4+2)$

12. $(8+17)/(7-2)$

11. ______________

12. ______________

13. $150\div(8+7)$

14. $72\div(5-2)\cdot 8$

13. ______________

14. ______________

15. $(27+63)\div(5\cdot 4-2)$

16. $(7+3)\cdot 15-24\div 3$

15. ______________

16. ______________

17. $17+3\cdot 6^2$

18. $4^3\cdot 3^2\div 12-5$

17. ______________

18. ______________

19. $(85-3\cdot 12)/7$

20. $225/(41-2^4)$

19. ______________

20. ______________

Name: Date:
Instructor: Section:

21. $(5^2 - 19)^2$

22. $8 \cdot 3^2 - 15 \cdot 4 + 6$

21. ______________

22. ______________

23. $7^2 \cdot 2^3 \div 14 - 5$

24. $7 \cdot 3^3 - 7^2 \cdot 3$

23. ______________

24. ______________

25. $74 - 3 \cdot (19 - 3 \cdot 4) + 5^0$

26. $7^2 + 3^3$

25. ______________

26. ______________

27. $(5^2 - 10)/3$

28. $(4 - 3)^0$

27. ______________

28. ______________

Concept Connections

29. Name the operations for which the Commutative Property does *not* hold.

____

____

For #10-11, use the area formula $A = b \cdot h \div 2$.

10. The base of a triangle is 9 feet and its height is 8 feet. Find the area of the triangle.

11. The base of a triangle is 38 cm and its height is 17 cm. Find the area of the triangle.

10. ______________

11. ______________

For #12-13, use the perimeter formula $P = a + b + c$.

12. The sides of a triangle are 5 feet, 8 feet and 11 feet. Find the perimeter of the triangle.

13. The sides of a triangle are 21 cm, 26 cm and 39 cm. Find the perimeter of the triangle.

12. ______________

13. ______________

For #14-15, a rectangle is 36 feet long and 28 feet wide.

14. What is the perimeter of the rectangle?

15. What is the area of the rectangle?

14. ______________

15. ______________

For #16-18, a rectangular driveway is 50 feet long and 9 feet wide. The driveway is to be coated with blacktop sealer. One can of sealer costs $18 and covers 225 square feet.

16. Determine the area of the driveway.

17. Determine the number of cans of blacktop sealer needed to cover the driveway.

16. ______________

17. ______________

Name: Date:
Instructor: Section:

18. Determine the cost of sealing the driveway.

18. ____________

For #19-20, use the formula $F = (9C \div 5) + 32$.

19. Convert 25° Celsius into degrees Fahrenheit.

20. Convert 15° Celsius into degrees Fahrenheit.

19. ____________

20. ____________

For #21-22, use the formula $C = 5(F - 32) \div 9$.

21. Convert 50° Fahrenheit into degrees Celsius.

22. Convert 104° Fahrenheit into degrees Celsius.

21. ____________

22. ____________

For #23-28, translate each verbal rule into a symbolic rule, using x as the input and y as the output.

23. The output is the sum of the input and 57.

24. The output is twenty less than the input.

23. ____________

24. ____________

25. The output is eight times the input.

26. The output is seven less than five times the input.

25. ____________

26. ____________

27. The output is four times the difference of the input and eleven.

28. The output is fifty less than the input squared.

27. ________________

28. ________________

Concept Connections

29. What is the difference between area and perimeter?

____

____

30. You and a friend are working on area and perimeter exercises. Your friend says the answer is 20 sq. ft. How would you be able to determine whether your friend was talking about an area or perimeter exercise?

____

____

Name: Date:
Instructor: Section:

Chapter 2 VARIABLES AND PROBLEM SOLVING

Activity 2.2

Learning Objectives

1. Recognize the input/output relationship between variables in a formula or equation (two variables only).
2. Evaluate variable expressions in formulas and equations.
3. Generate a table of input and corresponding output values from a given equation, formula, or situation.
4. Read, interpret, and plot points in rectangular coordinates that are obtained from evaluating a formula or equation.

Key Terms

Use the vocabulary terms listed below to complete each statement in Exercises 1–5.

horizontal	**vertical**	**graph**
input	**output**	**ordered pair**

1. When viewing a table of paired data values, the ________________ is the value that is considered first and the ________________ results from the first value.

2. The output variable is referenced on the ________________ axis.

3. An input value and its corresponding output value can be written as a(n) ________________ of numbers within a set of parentheses, separated by a comma.

4. The input variable is referenced on the ________________ axis.

5. The collection of all the points determined by a formula results in a curve called the ________________ of the formula or equation.

Practice Exercises

For #6-8, use the formula $y = 5x + 2$.

6. Complete the table of values for this formula. **6.** ________________

x	0	1	2	3	4
y					

7. Which variable is the input variable? **7.** ________________

8. Which variable is the output variable? **8.** ________________

9. Using the formula $F = (9C \div 5) + 32$, complete the table of values.

C degrees Celsius	F degrees Fahrenheit
0	
5	
10	
15	
20	

9. ______________

10. Using the formula $C = 5(F - 32) \div 9$, complete the table of values.

F degrees Fahrenheit	C degrees Celsius
104	
140	
176	
212	

10. ______________

For #11-12, use the formula $T = 60S$.

11. Which letter represents the input variable?

12. Which letter represents the output variable?

11. ______________

12. ______________

For #13-14, use the formula $B = 7x^2 + 5x - 3$.

13. Which letter represents the input variable?

14. Which letter represents the output variable?

13. ______________

14. ______________

Name: Date:
Instructor: Section:

For #15-16, use the formula $P = 4s$.

15. Which letter represents the input variable?

16. Which letter represents the output variable?

15. ______________

16. ______________

For #17-22, the formula $h = 112t - 16t^2$ *gives the height above the ground at time t, for a rocket that is fired straight up in the air with a velocity of 112 feet per second.*

17. Complete a table of values for the following input values: $t = 0, 1, 2, 3, 4, 5, 6, 7$ seconds.

17. ______________

18. Plot the input/output pairs from the table from Exercise #17 on a coordinate system and connect them in a smooth curve.

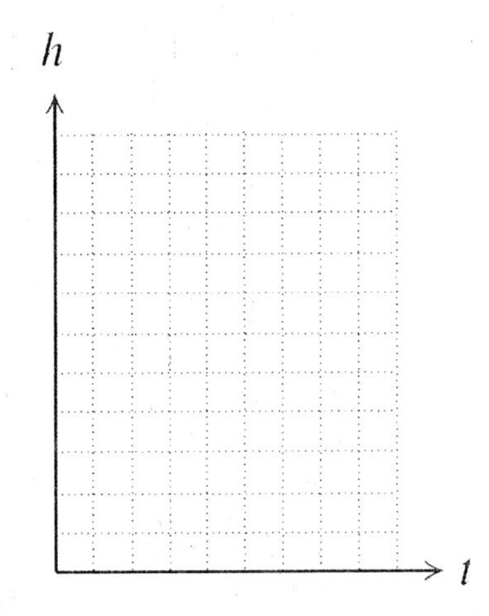

19. Use the graph to estimate how high the rocket goes.

20. From the graph when will the rocket be 196 ft above the ground?

19. ______________

20. ______________

21. When will the rocket hit the ground?

22. Give a reason for your answer in Exercise #21.

21. ______________

22. ______________

For #23-28, the formula $d = 65t$ gives the distance, in miles, that a car travels in t hours at a steady speed of 65 miles per hour.

23. Complete a table of values for the following input values: $t = 0, 1, 2, 3, 4$ hours.

23. ________________

24. Plot the input/output pairs from the table from Exercise #23 on a coordinate system and connect them in a smooth curve.

d

t

25. How long does it take to travel 195 miles?

26. What is the distance for 6 hours of travel?

25. ________________

26. ________________

27. How long does it take to travel 260 miles?

28. What is the distance for 8 hours of travel?

27. ________________

28. ________________

Concept Connections

29. From the formula before Exercise #17, evaluate the formula when $t = 9$. Interpret your answer.

____

____

30. From the table you created in Exercise #17, at what time is the rocket at 160 ft? Why do you have two answers?

____

____

Name: Date:
Instructor: Section:

Chapter 2 VARIABLES AND PROBLEM SOLVING

Activity 2.3

Learning Objectives

1. Translate contextual situations and verbal statements into equations.
2. Apply the fundamental principle of equality to solve equations of the forms $x + a = b$, $a + x = b$ and $x - a = b$.

Key Terms

Use the vocabulary terms listed below to complete each statement in Exercises 1–4.

equation **evaluated** **solved** **fundamental principle of equality**

1. The ________________ states that performing the same operation on both sides of a true equation will result in an equation that is also true.

2. A(n) ________________ is a statement that two expressions are equal.

3. An equation is ________________ when the unknown quantity is determined.

4. An expression is ________________ when the variables are replaced with known quantities and then the expression is simplified.

Practice Exercises

For #5-28, use an algebraic approach to solve each equation.

5. $x + 39 = 173$

6. $369 + s = 672$

7. $w - 59 = 111$

8. $359 = y + 147$

5. ________________

6. ________________

7. ________________

8. ________________

9. $t + 316 = 316$

10. $51 = y - 16$

11. $734 + x = 821$

12. $z - 58 = 232$

13. $w + 5670 = 8539$

14. $y - 562 = 196$

15. $5600 + x = 6120$

16. $359 = t - 360$

17. $y + 991 = 991$

18. $62,400 = z + 48,320$

19. $x - 727 = 0$

20. $x - 58 = 118$

9. ________________

10. ________________

11. ________________

12. ________________

13. ________________

14. ________________

15. ________________

16. ________________

17. ________________

18. ________________

19. ________________

20. ________________

Name: Date:
Instructor: Section:

21. $916 + s = 1123$

22. $w - 51 = 210$

21. ________________

22. ________________

23. $73 + x = 93$

24. $518 = t - 519$

23. ________________

24. ________________

25. $m + 88 = 88$

26. $150 + x = 151$

25. ________________

26. ________________

27. $y - 362 = 0$

28. $y - 35 = 17$

27. ________________

28. ________________

Concept Connections

29. In your own words state the difference between an expression and an equation.

____

____

30. Why are addition and subtraction called inverse operations?

____

____

Name: Date:
Instructor: Section:

Chapter 2 VARIABLES AND PROBLEM SOLVING

Activity 2.4

Learning Objectives

1. Apply the fundamental principle of equality to solve equations in the form $ax = b$, $a \neq 0$.
2. Translate contextual situations and verbal statements into equations.
3. Use the relationship rate · time = amount in various contexts.

Practice Exercises

For #1-28, use the fundamental principle of equality to solve each equation.

1. $5x = 45$

2. $13t = 195$

3. $350 = 7y$

4. $16w = 592$

5. $196 = 14z$

6. $3w = 171$

7. $389w = 778$

8. $3y = 5022$

1. ______________

2. ______________

3. ______________

4. ______________

5. ______________

6. ______________

7. ______________

8. ______________

9. $673t = 0$

10. $624 = 4x$

11. $999 = 27x$

12. $9y = 387$

13. $37 + x = 59$

14. $15v = 600$

15. $w - 37 = 202$

16. $1280 = 16y$

17. $s + 36 = 51$

18. $66 = p - 16$

19. $425g = 1275$

20. $0 = d - 66$

9. ______________

10. ______________

11. ______________

12. ______________

13. ______________

14. ______________

15. ______________

16. ______________

17. ______________

18. ______________

19. ______________

20. ______________

Name: Date:
Instructor: Section:

21. $18r = 684$

22. $4967 = 2489 + x$

21. ________________

22. ________________

23. $962 = z - 423$

24. $23w = 621$

23. ________________

24. ________________

25. $32,574 = 6t$

26. $x + 527 = 982$

25. ________________

26. ________________

27. $45 = w + 33$

28. $18y = 558$

27. ________________

28. ________________

Concept Connections

29. Given an equation in the general form $ax = b$, it is stated that a is a nonzero number. Explain why is a not zero.

____

____

30. Besides addition and subtraction, name two operations that are inverse operations. Explain why.

____

____

Name: Date:
Instructor: Section:

Chapter 2 VARIABLES AND PROBLEM SOLVING

Activity 2.5

Learning Objectives
1. Identify like terms.
2. Combine like terms using the distributive property.
3. Solve equations of the form $ax + bx = c$.

Practice Exercises

For #1-4, use the algebraic expression $6x + 4y + 7x$.

1. How many terms are in the expression?

2. What are the coefficients?

3. What are the like terms?

4. Simplify the expression by combining like terms.

1. ______________
2. ______________
3. ______________
4. ______________

For #5-7, use the algebraic expression $8a + 6b + 8a + 8b$.

5. How many terms are in the expression?

6. What are the coefficients?

7. What are the like terms?

5. ______________
6. ______________
7. ______________

For #8-13, combine like terms and rewrite each expression.

8. $13x-7x+5y+9y$ **9.** $3s+21t+8s+5t+25$

8. ______________

9. ______________

10. $11x+9x+3x-4y$ **11.** $6b+13c+c+8b-17$

10. ______________

11. ______________

12. $37p+49n+13p+15n+2n$

12. ______________

13. $153w-127w+150z+35z+144y-66y$

13. ______________

For #14-28, solve each equation. Check your answers in the original equation.

14. $7x+12x=570$ **15.** $15y-6y=180$

14. ______________

15. ______________

16. $385=22s-15s$ **17.** $0=69w+81w$

16. ______________

17. ______________

Name: Date:
Instructor: Section:

18. $6x + 13x - 8x = 176$

19. $34 = 5t - 2t + 14t$

20. $y + 5y - 2y = 176$

21. $16x + 26x - 35x = 182$

22. $2805 = 51t + 42t - 88t$

23. $23x - 21x + 17 = 95$

24. $54 = 19x - 18x + 7$

25. $68y - 73y + 67 + 8y = 88$

26. $x + 9x + 11x - 20x = 110$

27. $21x + 25x - x = 1260$

18. ______________

19. ______________

20. ______________

21. ______________

22. ______________

23. ______________

24. ______________

25. ______________

26. ______________

27. ______________

28. $4w + 3w + 6w - 5w + w - 7w + w = 111$ **28.** ______________

Concept Connections

29. When solving equations, how do you check your answer?

____

____

30. Why is it important to check your answer?

____

____

Name: Date:
Instructor: Section:

Chapter 2 VARIABLES AND PROBLEM SOLVING

Activity 2.6

Learning Objectives
1. Use the basic steps for problem solving.
2. Translate verbal statements into algebraic equations.
3. Use the basic principles of algebra to solve real-world problems.

Practice Exercises

For #1-23, translate each statement into an equation and then solve the equation for the unknown number.

1. An unknown number plus 361 is equal to 504.

2. The product of a number and 42 is 2310.

1. ____________

2. ____________

3. A number minus 627 is 132.

4. The sum of an unknown number and 732 is 3649.

3. ____________

4. ____________

5. Sixty subtracted from a number is 32.

6. The prime factors of 111 are 37 and an unknown number.

5. ____________

6. ____________

7. 19 plus an unknown number is 43.

8. The difference between a number and 67 is 132.

9. 791 is the product of 7 and an unknown number.

10. The sum of an unknown number and itself is 862.

11. An unknown number plus 669 is equal to 883.

12. The product of some number and 32 is 2400.

13. 228 is the product of 38 and a number.

14. What number times 9 equals 153?

7. ______________

8. ______________

9. ______________

10. ______________

11. ______________

12. ______________

13. ______________

14. ______________

Name: Date:
Instructor: Section:

15. The sum of an unknown number and 532 is 3607.

16. Nineteen subtracted from a number is 63.

15. ______________

16. ______________

17. The prime factors of 689 are 53 and an unknown number.

18. The sum of an unknown number and itself is 388.

17. ______________

18. ______________

19. The difference between a number and 51 is 23.

20. A number minus 542 is 876.

19. ______________

20. ______________

21. 833 is the product of 7 and an unknown number.

22. The product of 52 and an unknown number is 1092.

21. ______________

22. ______________

23. The sum of 97 and a number is 112.

23. ______________

For #24-28, solve each problem by applying the four steps of problem solving. Use the strategy of solving an algebraic equation for each problem.

24. For work, you need to rent a pickup truck. The truck can be rented for $85 per day with unlimited mileage. If your budget for the job is $425 for truck rental, how many days can you use the truck?

24. ________________

25. You plan to drive 455 miles to a conference. How fast must you drive to be there in 7 hours?

25. ________________

26. A rectangle that has an area of 437 square centimeters is 19 cm wide. How long is the rectangle?

26. ________________

27. Your goal is to save $1800 to pay for next year's books and fees. How much must you save each month if you have 8 months to accomplish your goal?

27. ________________

Name: Date:
Instructor: Section:

28. A rectangular field is 4 times longer than it is wide. If the perimeter is 650 feet, what are the dimensions (length and width) of the field?

28. ______________

Concept Connections

29. State The Four Steps of Problem Solving.

____

____

30. When solving a real-world problem, you found the value for the variable. How do you know when you're done with the problem?

____

____

Name: Date:
Instructor: Section:

Chapter 3 PROBLEM SOLVING WITH INTEGERS

Activity 3.1

Learning Objectives
1. Recognize integers.
2. Represent quantities in real-world situations using integers.
3. Represent integers on the number line.
4. Compare integers.
5. Calculate absolute values of integers.

Key Terms
Use the vocabulary terms listed below to complete each statement in Exercises 1–2.

positive **negative** **left** **right**

1. ____________________ numbers are numbers less than zero and ____________________ numbers are numbers greater than zero.

2. If $a > b$, then a is to the ____________________ of b on a number line. If $a < b$, then a is to the ____________________ of b on a number line.

Practice Exercises
For #3-8, write < or > between each number to make the statement true.

3. 11 □ 9

4. 0 □ 17

5. −6 □ −3

6. −9 □ 0

7. −4 □ −20

8. 6 □ −5

3. ____________

4. ____________

5. ____________

6. ____________

7. ____________

8. ____________

For #9-14, determine the value of each absolute value.

9. $|6|$ **10.** $|-23|$

11. $|59|$ **12.** $|-4|$

13. $|-100|$ **14.** $|0|$

9. ______________

10. ______________

11. ______________

12. ______________

13. ______________

14. ______________

For #15-20, determine the opposite of each number.

15. opposite of –11 **16.** opposite of 65

17. opposite of 3 **18.** opposite of –37

19. opposite of $|-43|$ **20.** opposite of $|62|$

15. ______________

16. ______________

17. ______________

18. ______________

19. ______________

20. ______________

Name: Date:
Instructor: Section:

For #21-28, express each quantity as an integer.

21. The stock price dropped by 55 points.

22. The scuba diver is 107 feet below the surface.

21. ______________

22. ______________

23. The Bears lost 10 yards on a penalty.

24. The temperature rose 17° at noon.

23. ______________

24. ______________

25. At sundown, the temperature dropped by 22°.

26. $25 is withdrawn from your checking account.

25. ______________

26. ______________

27. $130 is deposited in your checking account.

28. The submarine is 250 feet below the surface.

27. ______________

28. ______________

Concept Connections

29. In your own words, give the definition of absolute value.

____

____

30. In your own words explain how two numbers are opposites.

____

____

Name: Date:
Instructor: Section:

Chapter 3 PROBLEM SOLVING WITH INTEGERS

Activity 3.2

Learning Objectives

1. Add and subtract integers.
2. Identify properties of addition and subtraction of integers.

Practice Exercises

For #1-20, perform the indicated operation.

1. $-4+(-11)$

2. $-7+3$

3. $5+(-17)$

4. $19-7$

5. $8-15$

6. $-3-11$

7. $-8+(-11)$

8. $(-8)+12$

1. ______________

2. ______________

3. ______________

4. ______________

5. ______________

6. ______________

7. ______________

8. ______________

9. $-7-11$

10. $-6-(-7)$

11. $0-(-9)$

12. $-5+0$

13. $18-18$

14. $18-(-18)$

15. $0+(-11)$

16. $-5+(-6)$

17. $-9-(-4)$

18. $-8-(-13)$

19. $7-10$

20. $(-11)-5$

9. ________________

10. ________________

11. ________________

12. ________________

13. ________________

14. ________________

15. ________________

16. ________________

17. ________________

18. ________________

19. ________________

20. ________________

Name: Date:
Instructor: Section:

For #21-28, perform the indicated operations.

21. $6+(-7)-11$

22. $-7-8+5$

23. $-8-4-9$

24. $5-(-5)-5$

25. $-3-(-8)+(-5)$

26. $11+8+(-6)$

27. $15-(-8)-15$

28. $2-6-(-3)$

21. ______________

22. ______________

23. ______________

24. ______________

25. ______________

26. ______________

27. ______________

28. ______________

Concept Connections

29. You look at a topological map of California. You notice that the highest point in California is Mt. Whitney at 14,494 ft and the lowest point is in Death Valley at –282 ft. What is the difference between the highest and lowest points in California?

____

____

30. State the associative property of addition. Create an example showing the associative property is not true in general for subtraction.

____

____

Name: Date:
Instructor: Section:

Chapter 3 PROBLEM SOLVING WITH INTEGERS

Activity 3.3

Learning Objectives

1. Write formulas from verbal statements.
2. Evaluate expressions in formulas.
3. Solve equations of the form $x + b = c$ and $b - x = c$.
4. Solve formulas for a given variable.

Practice Exercises

For #1-10, solve each formula for the given variable.

1. $B = A + 314$, for A

2. $D = R - S$, for S

3. $P = R - C$, for R

4. $Q = R + S$, for S

5. $B = A - C$, for A

6. $M - P = L$, for P

7. $C = A + 54$, for A

8. $A - B = C$, for B

1. ________________

2. ________________

3. ________________

4. ________________

5. ________________

6. ________________

7. ________________

8. ________________

9. $F = C + D$, for C

10. $54 + D = A$, for D

9. ________________

10. ________________

For #11-20, let x represent an integer. Translate the verbal expression into an equation and solve for x.

11. The sum of an integer and 12 is –15.

12. The sum of an integer and –17 is 23.

11. ________________

12. ________________

13. 10 less than an integer is 25.

14. 16 less than an integer is –35.

13. ________________

14. ________________

15. The result of subtracting 29 from an integer is –6.

16. The result of subtracting 81 from an integer is –95.

15. ________________

16. ________________

17. The result of subtracting 35 from an integer is 5.

18. If an integer is subtracted from 11, the result is 8.

17. ________________

18. ________________

19. If an integer is subtracted from 3, the result is 9.

20. If an integer is subtracted from 5, the result is 5.

19. ________________

20. ________________

Name: Date:
Instructor: Section:

For #21-28, substitute the given values into the equation $c = a - b$ and determine the remaining variable.

21. $a = 15,\ b = -29$

22. $b = 11,\ c = -62$

21. ______________

22. ______________

23. $c = 76,\ b = 59$

24. $a = -23,\ b = 32$

23. ______________

24. ______________

25. $a = 54,\ b = -2$

26. $c = -41,\ a = 5$

25. ______________

26. ______________

27. $b = 21,\ c = -10$

28. $a = 4,\ c = 3$

27. ______________

28. ______________

Concept Connections

29. Explain why the formula $x = y + 2b - x$ is not solved for x.

____

____

30. After receiving your federal refund, you decide to splurge and buy a new TV. You spend $566 which includes $46 in taxes. What was the price of the TV before taxes?

____

____

Name:	Date:
Instructor:	Section:

Chapter 3 PROBLEM SOLVING WITH INTEGERS

Activity 3.4

Learning Objectives

1. Translate verbal rules into equations.
2. Determine an equation from a table of values.
3. Use a rectangular coordinate system to represent an equation graphically.

Key Terms

Use the vocabulary terms listed below to complete each statement in Exercises 1–3.

origin **quadrants** **scaling**

1. The two perpendicular coordinate axes divide the plane into four ________________.

2. The process of determining and labeling an appropriate distance between tick marks is called ________________.

3. The point of intersection of the axes is called the ________________ and has coordinates (0, 0).

Practice Exercises

4. Plot these points. $(2,\ 4)$ $(-3,\ 1)$ $(2,-5)$ $(-4,-1)$ $(0,-2)$ $(2,\ 0)$ $(0,\ 3)$ $(-4,\ 0)$

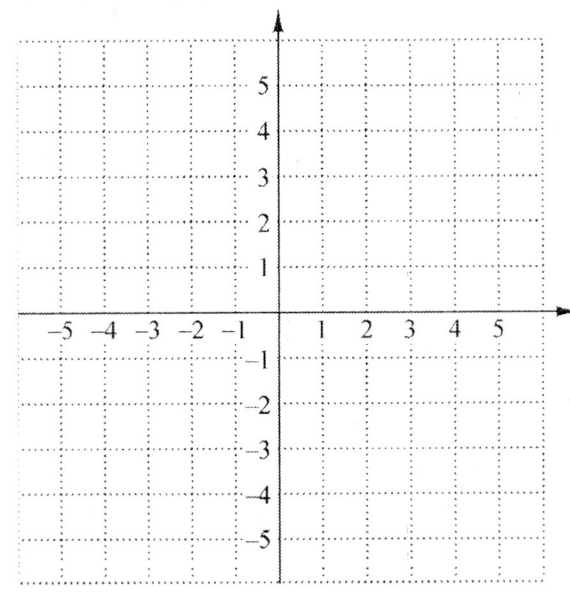

For #5-19, name the quadrant or axis where each point lies.

5. $(-2, 5)$

6. $(4, 0)$

7. $(3, -1)$

8. $(6, 6)$

9. $(0, -2)$

10. $(-1, -1)$

11. $(-5, 0)$

12. $(5, 4)$

13. $(-4, 2)$

14. $(-3, -5)$

15. $(1, -3)$

16. $(0, 5)$

5. ________________

6. ________________

7. ________________

8. ________________

9. ________________

10. ________________

11. ________________

12. ________________

13. ________________

14. ________________

15. ________________

16. ________________

Name: Date:
Instructor: Section:

17. $(-1, -3)$

18. $(1, -1)$

17. ________________

18. ________________

19. $(3, 0)$

19. ________________

For #20-24, find the coordinates of the points A, B, C, D, and E.

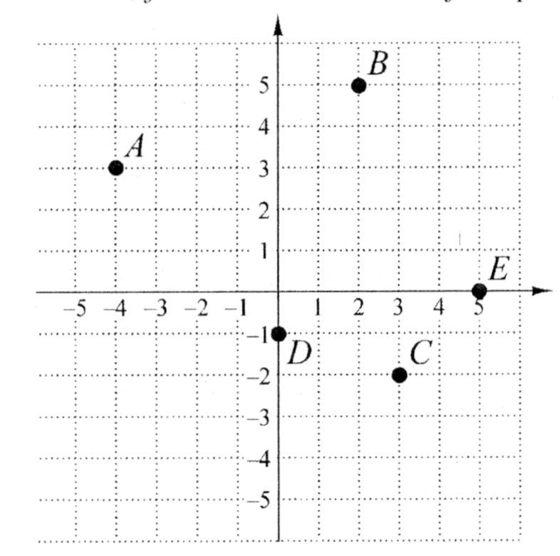

20. A________________

21. B________________

22. C________________

23. D________________

24. E________________

For #25-26, the difference of two integers is 8.

25. Translate the verbal rule into an equation.

25. ________________

26. Use the equation from Exercise #25 to complete the following table.

x	y
8	
	-2
4	
	-6
0	

26. ________________

For #27-28, the sum of two integers is –2.

27. Translate the verbal rule into an equation. **27.** ______________

28. Use the equation from Exercise #27 to complete the following table. **28.** ______________

x	y
5	
	-3
-3	
	0
-1	

Concept Connections

29. Draw a rectangular coordinate system that has proper scaling and plot the points (0, 0), (1, 1), (2, 2), (3, 3), (4, 4) and (5, 5). Next draw another rectangular coordinate system that does NOT have equal spaces between tick marks, and plot the same points. Compare the two graphs.

____

____

30. Sue can't remember which value goes first when plotting points. What verbal rule can you state that explains which coordinate to plot first and its location on the graph?

____

____

Name: Date:
Instructor: Section:

Chapter 3 PROBLEM SOLVING WITH INTEGERS

Activity 3.5

Learning Objectives

1. Multiply and divide integers.
2. Perform calculations that involve a sequence of operations.
3. Apply exponents to integers.
4. Identify properties of calculations that involve multiplication and division with zero.

Practice Exercises

For #1-28, determine the product or quotient.

1. $-6(-8)$

2. $(-9)(-7)$

3. $-7 \cdot 3$

4. $(-5)(-8)$

5. $-11(0)$

6. $5(-6)$

7. $(-4)(-7)$

8. $63 \div (-9)$

1. ____________

2. ____________

3. ____________

4. ____________

5. ____________

6. ____________

7. ____________

8. ____________

9. $-35 \div (-7)$

10. $0 \div (-6)$

11. $-14 \div 0$

12. $-24 \div 6$

13. $18 \div (-3)$

14. $-54 \div 9$

15. $\frac{-72}{-1}$

16. $\frac{0}{-31}$

17. $2(-11)(-5)$

18. $(-7)(-8)(-1)(4)(3)$

19. $-27 \div 0$

20. $0 \div (-50)$

9. ________________

10. ________________

11. ________________

12. ________________

13. ________________

14. ________________

15. ________________

16. ________________

17. ________________

18. ________________

19. ________________

20. ________________

Name: Date:
Instructor: Section:

21. $(-8)^2$

22. -8^2

23. $(-1)^6$

24. 5^3

25. $(13)^2$

26. $-(13)^2$

27. $(-5)(-1)(-2)(-3)$

28. $-(-5)^3$

21. ____________

22. ____________

23. ____________

24. ____________

25. ____________

26. ____________

27. ____________

28. ____________

Concept Connections

29. After a series of snow storms in the winter, the snow pack level in the Sierra mountains was measured at 38 ft. Over the course of several weeks, additional readings were taken.

Week	1	2	3	4	5
Change in snow pack (feet)	-3	2	-5	1	-8

Determine the total change in the snow pack level after week 5.
Determine the snow pack level after week 5.

____

____

30. Do $(-x)^2$ and $-x^2$ represent equivalent quantities? Why or why not?

____

____

Name: Date:
Instructor: Section:

Chapter 3 PROBLEM SOLVING WITH INTEGERS

Activity 3.6

Learning Objectives

1. Use order of operations with expressions that involve integers.
2. Apply the distributive property.
3. Evaluate algebraic expression and formulas using integers.
4. Combine like terms.
5. Solve equations of the form $ax = b$, where $a \neq 0$, that involve integers.
6. Solve equations of the form $ax + bx = c$, where $a + b \neq 0$, that involve integers.

Practice Exercises

For #1-10, evaluate each expression.

1. $(7-15) \div 4 + 9$

2. $-6+5(2-7)$

3. $-4^2 \cdot 3^2 + 64$

4. $-9+12 \div (8-10)$

5. $(3-8)5$

6. $(3+4)^2 - 25$

7. $-3+(3 \cdot 5 - 4 \cdot 6)$

8. $(3 \cdot 4 - 5 \cdot 6) \div (-9)$

1. ________________

2. ________________

3. ________________

4. ________________

5. ________________

6. ________________

7. ________________

8. ________________

9. $-7+21\div(16-37)$

10. $4^2(-6)^2\div(-3)$

9. ______________

10. ______________

For #11-13, use the distributive property to evaluate.

11. $7(-5+6)$

12. $-5(7-9)$

11. ______________

12. ______________

13. $8(-9-5)$

13. ______________

For #14-16, evaluate each expression using the given values.

14. x^2y+5z, for $x=-4$, $y=-7$, $z=57$

14. ______________

15. $(7a+4b)(3a-b)$, for $a=-5$, $b=8$

15. ______________

16. $\dfrac{9bc-7a}{ab}$, for $a=-3$, $b=-7$, $c=-5$

16. ______________

Name: Date:
Instructor: Section:

For #17-20, solve each equation.

17. $6x = 42$

18. $11s = -110$

17. ______________

18. ______________

19. $-18 = -6y$

20. $-x = -8$

19. ______________

20. ______________

For #21-24, combine like terms.

21. $a - 3b + 7a + 15b$

21. ______________

22. $13x + 14y - 17x - 8y + x$

22. ______________

23. $8x - 7y + x - 3y + 12y - 3x$

23. ______________

24. $2b - a + b - 3a - 6b$

24. ______________

For #25-28, solve each equation by first combining like terms.

25. $17x - 23x = -48$

26. $-20y - 8y = 140$

25. ______________

26. ______________

27. $-18y + 13y = 55$

28. $8x - 11x = 51$

27. ______________

28. ______________

Concept Connections

29. After a series of rainstorms, a river crested to 30 ft, near flood stage. After 6 days, the river was marked at 24 ft. What was the daily average change of the river over the 6-day period?

____

____

30. The order of operations was first introduced in Chapter 1 with whole numbers. Do the same rules for the order of operations still apply to integers? If so, then explain the order of operations for arithmetic expressions containing parentheses, addition, subtraction, multiplication, division and exponentiation.

____

____

____

Name: Date:
Instructor: Section:

Chapter 4 PROBLEM SOLVING WITH FRACTIONS

Activity 4.1

Learning Objectives

1. Identify the numerator and the denominator of a fraction.
2. Determine the greatest common factor (GCF).
3. Determine equivalent fractions.
4. Reduce fractions to equivalent fractions in lowest terms.
5. Determine the least common denominator (LCD) of two or more fractions.
6. Compare fractions.

Key Terms

Use the vocabulary terms listed below to complete each statement in Exercises 1–6.

greatest common factor	**equivalent fraction**	**rational number**
lowest terms	**numerator**	**denominator**

1. A ____________________ is a number that can be written in the form $\frac{a}{b}$, where a and b are integers and b is not zero.

2. The bottom number of a fraction is called the ____________________ , which indicates the total number of equal parts into which a whole unit is divided.

3. If the largest factor of both the numerator and denominator of a fraction is 1, the fraction is said to be in ____________________ .

4. The top portion of a fraction is called the ____________________ , which indicates the number of equal parts that the fraction represents out of the total number of parts.

5. The ____________________ of two or more numbers is the largest number that is a factor of each of the given numbers.

6. To write a(n) ____________________ , multiply or divide both the numerator and denominator of a given fraction by the same number.

Practice Exercises

For #7-14, determine the missing number.

7. $\frac{5}{7}=\frac{?}{28}$

8. $\frac{3}{5}=\frac{?}{20}$

7. ____________________

8. ____________________

9. $\frac{7}{9}=\frac{?}{81}$

10. $\frac{9}{10}=\frac{?}{200}$

9. ______________

10. ______________

11. $\frac{3}{11}=\frac{?}{88}$

12. $\frac{7}{9}=\frac{?}{54}$

11. ______________

12. ______________

13. $\frac{3}{40}=\frac{?}{120}$

14. $\frac{8}{13}=\frac{?}{195}$

13. ______________

14. ______________

For #15-18, reduce each fraction to an equivalent fraction in lowest terms.

15. $\frac{7}{35}$

16. $\frac{8}{56}$

15. ______________

16. ______________

17. $\frac{12}{39}$

18. $\frac{9}{51}$

17. ______________

18. ______________

Name: Date:
Instructor: Section:

For #19-24, use > or < to compare the fractions.

19. $\frac{6}{11}$ and $\frac{1}{2}$

20. $\frac{6}{7}$ and $\frac{3}{4}$

21. $\frac{3}{5}$ and $\frac{5}{9}$

22. $\frac{2}{3}$ and $\frac{5}{7}$

23. $\frac{4}{7}$ and $\frac{4}{5}$

24. $\frac{3}{8}$ and $\frac{2}{5}$

19. ____________

20. ____________

21. ____________

22. ____________

23. ____________

24. ____________

For #25-28, determine the least common denominator for each set of fractions.

25. $\frac{7}{10}$ and $\frac{4}{15}$

26. $\frac{7}{12}$ and $\frac{5}{8}$

27. $\frac{7}{8}$ and $\frac{5}{18}$

28. $\frac{4}{25}$ and $\frac{5}{30}$

25. ____________

26. ____________

27. ____________

28. ____________

Concept Connections

29. In the definition of a rational number $\frac{a}{b}$, it is stated that b is not zero. Why?

____

____

30. Cynthia says that she can always find a common denominator of two fractions by multiplying the denominators together. How can you convince her that finding the LCD is more efficient when adding two fractions with different denominators?

____

____

Name: Date:
Instructor: Section:

Chapter 4 PROBLEM SOLVING WITH FRACTIONS

Activity 4.2

Learning Objectives
1. Multiply and divide fractions.
2. Recognize the sign of a fraction.
3. Determine the reciprocal of a fraction.
4. Solve equations of the form $ax = b$, $a \neq 0$, that involve fractions.

Key Terms
Use the vocabulary terms listed below to complete each statement in Exercises 1–5.

reciprocals **dividing** **multiplying** **zero** **one**

1. The reciprocal of one is ____________________ .

2. ____________________ does not have a reciprocal.

3. When ____________________ two fractions, you do not need to find the reciprocal of the second fraction.

4. Two numbers are ____________________ if their product is 1.

5. When ____________________ two fractions, you need to find the reciprocal of the second fraction.

Practice Exercises
For #6-12, multiply the fractions and express each answer in simplest form.

6. $\frac{5}{9} \cdot \frac{4}{7}$

7. $\frac{7}{8} \cdot \frac{3}{5}$

8. $\frac{-6}{13} \cdot \frac{39}{-56}$

9. $\frac{-5}{12} \cdot \frac{3}{-40}$

6. ____________________

7. ____________________

8. ____________________

9. ____________________

10. $\frac{18}{35}\cdot\frac{-14}{27}$

11. $\frac{13}{34}\cdot\frac{17}{26}$

10. ________________

11. ________________

12. $\frac{16}{35}\cdot\frac{-28}{64}$

12. ________________

For #13-19, divide the fractions and express each answer in simplest form.

13. $\frac{1}{6}\div\frac{1}{12}$

14. $-\frac{4}{9}\div\frac{16}{-27}$

13. ________________

14. ________________

15. $-7\div\frac{7}{11}$

16. $-\frac{13}{24}\div\frac{39}{8}$

15. ________________

16. ________________

17. $\dfrac{\frac{29}{15}}{-58}$

18. $\dfrac{-\frac{4}{9}}{-\frac{16}{27}}$

17. ________________

18. ________________

19. $\frac{5}{14}\div\frac{-15}{7}$

19. ________________

Name: Date:
Instructor: Section:

For #20-25, solve each equation.

20. $\frac{13}{15}x = \frac{26}{5}$

21. $-\frac{5}{12}x = \frac{7}{18}$

20. ____________

21. ____________

22. $\frac{2}{7}w = -\frac{5}{21}$

23. $-\frac{12}{35}y = \frac{3}{28}$

22. ____________

23. ____________

24. $\frac{17}{27}x = \frac{34}{135}$

25. $-\frac{5}{16}t = -\frac{15}{8}$

24. ____________

25. ____________

For #26-28, translate each sentence to an equation and solve.

26. The product of a number and 4 is $\frac{8}{5}$.

26. ____________

27. $\frac{3}{4}$ times a number is –6.

27. ____________

28. A number divided by 8 is 7.

28. ____________

Concept Connections

29. Why is it important to know how to find a reciprocal when dividing two fractions?

30. What is the difference between multiplying two fractions and dividing two fractions?

Name: Date:
Instructor: Section:

Chapter 4 PROBLEM SOLVING WITH FRACTIONS

Activity 4.3

Learning Objectives

1. Add and subtract fractions with the same denominator.
2. Add and subtract fractions with different denominators.
3. Solve equations in the form $x + b = c$ and $x - b = c$ that involve fractions.

Practice Exercises

For #1-18, perform the operation. Write the result in lowest terms.

1. $\frac{7}{20}+\frac{7}{20}$

2. $\frac{3}{4}-\frac{1}{4}$

1. ______________

2. ______________

3. $\frac{7}{12}+\frac{1}{12}$

4. $\frac{7}{8}-\frac{5}{8}$

3. ______________

4. ______________

5. $\frac{6}{17}-\frac{6}{17}$

6. $\frac{7}{15}-\frac{4}{15}$

5. ______________

6. ______________

7. $\frac{1}{25}+\frac{9}{25}$

8. $-\frac{3}{8}-\frac{1}{8}$

7. ______________

8. ______________

9. $-\frac{7}{16}-\frac{7}{16}$

10. $-\frac{5}{9}+\frac{2}{9}$

9. ________________

10. ________________

11. $\frac{3}{4}-\frac{3}{8}$

12. $\frac{1}{14}+\frac{3}{7}$

11. ________________

12. ________________

13. $-\frac{1}{18}-\frac{1}{9}$

14. $-\frac{3}{10}+\left(-\frac{1}{5}\right)$

13. ________________

14. ________________

15. $-\frac{1}{9}+\frac{1}{15}$

16. $-\frac{4}{7}+\frac{4}{21}+\frac{1}{3}$

15. ________________

16. ________________

17. $-\frac{5}{6}+\frac{3}{8}+\frac{1}{2}$

18. $\frac{2}{3}-\frac{5}{21}$

17. ________________

18. ________________

Name: Date:
Instructor: Section:

For #19-27, solve each equation.

19. $\frac{3}{7}=c-\frac{1}{14}$

20. $\frac{4}{9}=\frac{1}{3}-x$

19. ______________

20. ______________

21. $x-\frac{4}{15}=\frac{2}{3}$

22. $\frac{1}{3}+x=\frac{3}{4}$

21. ______________

22. ______________

23. $\frac{2}{3}+x=-\frac{1}{5}$

24. $-\frac{5}{13}=-\frac{3}{26}-y$

23. ______________

24. ______________

25. $\frac{9}{16}+b=-\frac{1}{8}$

26. $\frac{1}{3}+x=\frac{1}{4}$

25. ______________

26. ______________

27. $w+\frac{2}{5}=\frac{3}{15}$

28. $\frac{4}{9}=x+\frac{1}{3}$

27. ______________

28. ______________

Concept Connections

29. Explain in your own words how to determine the least common denominator (LCD) of fractions.

____

____

30. Explain in your own words how to add or subtract fractions with different denominators.

____

____

Name: Date:
Instructor: Section:

Chapter 4 PROBLEM SOLVING WITH FRACTIONS

Activity 4.4

Learning Objectives
1. Calculate powers and square roots of fractions.
2. Evaluate equations that involve powers.
3. Evaluate equations that involve square roots.
4. Use order of operations to calculate numerical expressions that involve fractions.
5. Evaluate algebraic expressions that involve fractions.
6. Use the distributive property with fractions.
7. Solve equations of the form $ax + bx = c$ with fraction coefficients.

Practice Exercises

For #1-10, evaluate and reduce to lowest terms.

1. $\left(\frac{3}{5}\right)^2$

2. $\left(-\frac{4}{5}\right)^3$

3. $\left(-\frac{5}{9}\right)^2$

4. $\left(-\frac{1}{5}\right)^4$

5. $\sqrt{\frac{81}{100}}$

6. $\sqrt{\frac{16}{25}}$

7. $\sqrt{\frac{75}{3}}$

8. $\sqrt{\frac{0}{7}}$

1. ______________

2. ______________

3. ______________

4. ______________

5. ______________

6. ______________

7. ______________

8. ______________

9. $\sqrt{\frac{49}{0}}$

10. $\sqrt{\frac{80}{5}}$

9. ______________

10. ______________

For #11-14, evaluate each expression.

11. $3x-6(x-5)$, where $x=\frac{1}{3}$

12. $x(x-5)+x(5x+3)$, where $x=\frac{1}{5}$

11. ______________

12. ______________

13. $5x^3+x^2$, where $x=\frac{1}{3}$

14. $5xy-3y^2$, where $x=\frac{4}{5}$ and $y=\frac{1}{3}$

13. ______________

14. ______________

For #15-26, solve each equation.

15. $\frac{7}{8}x=\frac{1}{5}$

16. $\frac{5}{13}x=-\frac{5}{17}$

15. ______________

16. ______________

17. $-\frac{7}{8}t=-\frac{7}{12}$

18. $-\frac{5}{6}y=\frac{25}{36}$

17. ______________

18. ______________

Name: Date:
Instructor: Section:

19. $-\frac{2}{3}y=-\frac{5}{12}$

20. $\frac{2}{7}x+\frac{1}{2}x=\frac{3}{14}$

19. ______________

20. ______________

21. $\frac{1}{12}x-\frac{5}{6}x=\frac{9}{16}$

22. $-\frac{1}{6}x+\frac{1}{2}x=\frac{5}{24}$

21. ______________

22. ______________

23. $-\frac{7}{9}x+\frac{5}{6}x=\frac{5}{9}$

24. $-\frac{1}{4}y-\frac{1}{6}y=\frac{7}{24}$

23. ______________

24. ______________

25. $\frac{3}{5}b+\frac{3}{4}b=-\frac{9}{10}$

26. $\frac{11}{18}x-\frac{7}{9}x=-\frac{7}{48}$

25. ______________

26. ______________

For #27-28, recall that the hang time t for a jump is given by $t=\frac{1}{2}\sqrt{s}$*, where s is the height of the jump in feet, and t is measured in seconds. Determine the hang time for the following heights.*

27. $s=4$ feet

28. $s=\frac{1}{3}$ yard

27. ______________

28. ______________

Concept Connections

29. In the procedure to raise a fraction to a power $\left(\frac{a}{b}\right)^n$, it is stated that $b\neq 0$. Why?

____

____

30. Explain in your own words how to determine the square root of a fraction.

____

____

Name: ,   Date:
Instructor: Section:

Chapter 5 PROBLEM SOLVING WITH MIXED NUMBERS AND DECIMALS

Activity 5.1

Learning Objectives

1. Determine equivalent fractions.
2. Add and subtract fractions and mixed numbers with the same denominator.
3. Convert mixed numbers to improper fractions and improper fractions to mixed numbers.

Key Terms

Use the vocabulary terms listed below to complete each statement in Exercises 1–2.

mixed number **improper fraction**

1. $\frac{13}{5}$ is an example of a(n) ____________________ , since the absolute value of the numerator is greater than or equal to the absolute value of the denominator..

2. $4\frac{5}{8}$ is an example of a(n) ____________________ , since it is the sum of an integer and a proper fraction.

Practice Exercises

For #3-10, reduce each improper fraction to lowest terms. Convert to a mixed number.

3. $\frac{6}{5}$ **4.** $-\frac{7}{5}$

5. $-\frac{9}{4}$ **6.** $\frac{10}{7}$

3. ________________

4. ________________

5. ________________

6. ________________

7. $-\frac{32}{24}$

8. $\frac{90}{81}$

7. ______________

8. ______________

9. $-\frac{63}{12}$

10. $\frac{48}{15}$

9. ______________

10. ______________

For #11-18, convert each mixed number to an improper fraction in lowest terms.

11. $5\frac{7}{8}$

12. $-3\frac{4}{5}$

11. ______________

12. ______________

13. $-6\frac{3}{4}$

14. $8\frac{8}{9}$

13. ______________

14. ______________

15. $-4\frac{8}{12}$

16. $-6\frac{10}{15}$

15. ______________

16. ______________

17. $7\frac{18}{24}$

18. $9\frac{25}{30}$

17. ______________

18. ______________

Name: Date:
Instructor: Section:

For #19-28, perform each operation. Write answer as a mixed number in simplest form.

19. $\frac{5}{7}+\frac{5}{7}$

20. $\frac{5}{6}+\frac{5}{6}$

21. $\frac{36}{24}-\frac{8}{24}$

22. $3\frac{2}{7}+5\frac{4}{7}$

23. $5\frac{5}{8}-3\frac{3}{8}$

24. $9\frac{2}{7}-4\frac{6}{7}$

25. $7\frac{10}{13}-9\frac{11}{13}$

26. $3\frac{11}{17}+\left(-10\frac{4}{17}\right)$

27. $-15\frac{7}{9}-\left(-8\frac{4}{9}\right)$

28. $-13\frac{17}{19}-11\frac{5}{19}$

19. ______________

20. ______________

21. ______________

22. ______________

23. ______________

24. ______________

25. ______________

26. ______________

27. ______________

28. ______________

Concept Connections

29. Show how the negative mixed number $-5\frac{8}{9}$, can be written as a sum of an integer and a proper fraction.

---

30. Create a general formula for converting from the mixed number $a\frac{b}{c}$ to an improper fraction.

Name: Date:
Instructor: Section:

Chapter 5 PROBLEM SOLVING WITH MIXED NUMBERS AND DECIMALS

Activity 5.2

Learning Objectives

1. Determine the least common denominator (LCD) for two or more mixed numbers.
2. Add and subtract mixed numbers with different denominators.
3. Solve equations of the form $x + b = c$ and $x - b = c$ that involve mixed numbers.

Practice Exercises

For #1-18, perform each operation.

1. $6\frac{1}{5}+8\frac{1}{10}$

2. $5\frac{3}{7}-4\frac{1}{14}$

3. $-2\frac{3}{4}+5\frac{7}{8}$

4. $7\frac{1}{5}+2\frac{1}{3}$

5. $9\frac{11}{18}+10\frac{14}{24}$

6. $14\frac{5}{6}+11\frac{4}{9}$

7. $6\frac{3}{18}+7\frac{9}{12}$

8. $14\frac{1}{6}-8\frac{1}{12}$

1. ______________

2. ______________

3. ______________

4. ______________

5. ______________

6. ______________

7. ______________

8. ______________

9. $20-11\frac{4}{5}$

10. $16\frac{2}{3}-9$

11. $14+8\frac{8}{9}$

12. $5\frac{5}{7}-3\frac{11}{14}$

13. $5\frac{3}{13}-3\frac{11}{26}$

14. $-8\frac{5}{9}+3\frac{5}{6}$

15. $-8\frac{2}{3}+\left(-5\frac{7}{15}\right)$

16. $3\frac{7}{10}-\left(-5\frac{6}{7}\right)$

17. $-8\frac{4}{5}+13\frac{7}{15}$

18. $7\frac{5}{8}-6$

9. ________________

10. ________________

11. ________________

12. ________________

13. ________________

14. ________________

15. ________________

16. ________________

17. ________________

18. ________________

Name: Date:
Instructor: Section:

For #19-28, solve for x.

19. $x-4\frac{2}{3}=7\frac{1}{2}$

20. $x+11=6\frac{3}{4}$

21. $x+4\frac{2}{5}=9$

22. $9\frac{5}{6}+x=4\frac{2}{3}$

23. $3\frac{3}{4}+x=7\frac{1}{6}$

24. $x-2\frac{7}{9}=8$

25. $x+7\frac{2}{3}=-4\frac{3}{4}$

26. $8\frac{1}{9}-x=7\frac{2}{3}$

27. $x-4\frac{3}{8}=2\frac{1}{2}$

28. $x-2\frac{3}{4}=6\frac{2}{9}$

19. ______________

20. ______________

21. ______________

22. ______________

23. ______________

24. ______________

25. ______________

26. ______________

27. ______________

28. ______________

Concept Connections

29. Alice incorrectly simplifies $2\frac{2}{3}+3\frac{7}{9}$ as $5\frac{9}{9}=6$. What did she do wrong? What is the correct solution?

---

30. Mike incorrectly simplifies $-7\frac{2}{9}+6\frac{7}{9}$ as $-1\frac{5}{9}$. What did he do wrong? What is the correct solution?

Name: Date:
Instructor: Section:

Chapter 5 PROBLEM SOLVING WITH MIXED NUMBERS AND DECIMALS

Activity 5.3

Learning Objectives
1. Multiply and divide mixed numbers.
2. Evaluate expressions with mixed numbers.
3. Calculate the square root of a mixed number.
4. Solve equations of the form $ax + b = 0$, $a \neq 0$, that involve mixed numbers.

Practice Exercises

For #1-6, multiply. Write the answer as a mixed number, if necessary.

1. $4\frac{1}{3} \cdot 5\frac{3}{5}$

2. $-6\frac{5}{12} \cdot 4\frac{1}{6}$

3. $\left(-8\frac{3}{13}\right) \cdot 3\frac{1}{3}$

4. $\left(-6\frac{5}{8}\right) \cdot (-16)$

5. $4\frac{4}{9} \cdot 3\frac{3}{5}$

6. $-3\frac{9}{10} \cdot 4\frac{4}{9}$

1. ________
2. ________
3. ________
4. ________
5. ________
6. ________

For #7-12, divide. Write the answer as a mixed number, if necessary.

7. $4\frac{4}{5} \div 1\frac{7}{9}$

8. $-2\frac{5}{6} \div \frac{6}{7}$

7. ________
8. ________

9. $5\frac{6}{7} \div 5\frac{1}{8}$

10. $\dfrac{-13\frac{3}{4}}{-7\frac{6}{7}}$

9. ________________

10. ________________

11. $-9 \div 3\frac{3}{5}$

12. $-8\frac{4}{5} \div \left(-6\frac{7}{8}\right)$

11. ________________

12. ________________

For #13-20, simplify each expression.

13. $\left(\frac{4}{5}\right)^2$

14. $\left(-\frac{5}{6}\right)^3$

13. ________________

14. ________________

15. $\left(-\frac{8}{9}\right)^2$

16. $\left(\frac{2}{5}\right)^4$

15. ________________

16. ________________

17. $\sqrt{\frac{25}{81}}$

18. $\sqrt{\frac{49}{64}}$

17. ________________

18. ________________

19. $x - 4\frac{2}{3} = 7\frac{1}{2}$

20. $x + 11 = 6\frac{3}{4}$

19. ________________

20. ________________

Name: Date:
Instructor: Section:

For #21-24, evaluate each expression.

21. $5x-3(x-2)$,
where $x=1\frac{2}{3}$

22. $4x+2(x-3)$,
where $x=\frac{3}{4}$

23. $5xy-x^2$, where
$x=\frac{1}{2}$ and $y=\frac{2}{5}$

24. x^2-x^3,
where $x=2\frac{2}{3}$

21. ______________

22. ______________

23. ______________

24. ______________

For #25-28, solve for x.

25. $3\frac{4}{5}x=38$

26. $5\frac{3}{5}x=-8\frac{1}{6}$

27. $6\frac{3}{4}x=-\frac{27}{32}$

28. $-4\frac{2}{3}x=-4\frac{1}{5}$

25. ______________

26. ______________

27. ______________

28. ______________

Concept Connections

29. The maximum distance, d, in kilometers that a person can see from a height h meters above the ground is given by the formula $d = \frac{7}{2}\sqrt{h}$. Tony works on the 17th floor in a skyscraper. Each floor of this building is $3\frac{7}{8}$ meters high. Tony's height is 2 m. What is the maximum distance Tony can see on a clear day? Note that the first floor is at 0 meters, the second floor at $3\frac{7}{8}$ meters, etc.

---

30. Julie incorrectly simplifies $2\frac{3}{4}\cdot 5\frac{5}{7}$ as $2\frac{3}{4}\cdot 5\frac{5}{7} = 10\frac{15}{28}$.
What did she do wrong? What is the correct solution?

Name:
Instructor:

Date:
Section:

Chapter 5 PROBLEM SOLVING WITH MIXED NUMBERS AND DECIMALS

Activity 5.4

Learning Objectives

1. Identify place values of numbers written in decimal form.
2. Convert a decimal to a fraction or a mixed number, and vice versa.
3. Classify decimals.
4. Compare decimals.
5. Read and write decimals.
6. Round decimals.

Practice Exercises

For #1-4, write each number in words.

1. 0.035

2. 0.00568

3. 560.0132

4. 7753.1559

1. ________________

____

2. ________________

____

3. ________________

____

4. ________________

____

For #5-8, write each as a decimal.

5. 73 ten-thousandths

6. two hundred eighty-two thousandths

7. six hundred forty-three and nine hundred eleven hundred-thousandths

8. three thousand eight hundred fifteen and eighty-one thousandths

5. ________________

6. ________________

7. ________________

8. ________________

For #9-12, round each number.

9. 0.08596 to the nearest tenth

10. 0.04878 to the nearest thousandth

9. ______________

10. ______________

11. 0.6989 to the nearest hundredth

12. 567.899186 to the nearest ten-thousandth

11. ______________

12. ______________

For #13-16, use the inequality symbols < or > to compare the decimals.

13. 56.3791 and 56.3789

14. 0.000604 and 0.000611

13. ______________

14. ______________

15. 67.19872 and 67.19869

16. 584.23811 and 584.239

15. ______________

16. ______________

For #17–28, write each fraction as a terminating decimal, or if a repeating decimal, round to the thousandth's place.

17. $\frac{17}{20}$

18. $\frac{35}{80}$

17. ______________

18. ______________

19. $\frac{24}{96}$

20. $\frac{27}{70}$

19. ______________

20. ______________

Name:
Instructor:

Date:
Section:

21. $\frac{42}{90}$

22. $\frac{39}{52}$

23. $\frac{5}{7}$

24. $\frac{8}{23}$

25. $\frac{9}{11}$

26. $4\frac{8}{29}$

27. $\frac{5}{2}$

28. $11\frac{7}{8}$

21. ________________

22. ________________

23. ________________

24. ________________

25. ________________

26. ________________

27. ________________

28. ________________

Concept Connections

29. What kind of numbers are included in the set of irrational numbers?

____

____

30. The Greek letter π, pi, represents a specific irrational number. One approximation of this number in fraction form is $\frac{22}{7}$ and another is $\frac{355}{113}$. Which fractional approximation of π is closer? Note that rounded to 8 decimal places $\pi \approx 3.14159265$.

____

____

Name: Date:
Instructor: Section:

Chapter 5 PROBLEM SOLVING WITH MIXED NUMBERS AND DECIMALS

Activity 5.5

Learning Objectives
1. Add and subtract decimals.
2. Compare and interpret decimals.
3. Solve equations of the type $x + b = c$ and $x - b = c$ that involve decimals.

Practice Exercises

For #1-16, perform each operation.

1. $15.7 + 2.98$ **2.** $8.304 + 8.821$

1. ________________

2. ________________

3. $2.67 - 2.734$ **4.** $0.376 + 0.115$

3. ________________

4. ________________

5. $3.345 - 1.016$ **6.** $-19.36 + 12.57$

5. ________________

6. ________________

7. $15.7 + 2.98$ **8.** $8.304 + 8.821$

7. ________________

8. ________________

9. $2.67 - 2.734$ **10.** $0.376 + 0.115$

9. ________________

10. ________________

11. $0.0081 - 0.00992$

12. $-0.563 + 0.403$

11. ______________

12. ______________

13. $-0.459 + 0.258$

14. $73.066 - 81.486$

13. ______________

14. ______________

15. $1010.10 - 1011.367$

16. $42.35 - 24.53$

15. ______________

16. ______________

For #17-28, solve each equation.

17. $2.05 + 0.386 + x = 7.8$

18. $0.002009 - x = 0.0084$

17. ______________

18. ______________

19. $x - 3.742 = 96.738$

20. $2.849 + y = 15.309$

19. ______________

20. ______________

21. $w + 16.36 = 10.5$

22. $0.6015 + x = 0$

21. ______________

22. ______________

Name: Date:
Instructor: Section:

23. $15.49 + b = 15.49$

24. $4.598 - x = 9.365$

23. ______________

24. ______________

25. $x + 62.87 = 453.2$

26. $p - 1.902 = 0.428$

25. ______________

26. ______________

27. $x - 26.517 = 32.416$

28. $b + 0.167 = 0.122$

27. ______________

28. ______________

Concept Connections

29. The freezing point of francium is 80.6°F. The boiling point is 1256°F. You have some francium stored at –20.4°F in a special container in your laboratory. What is the difference in temperature between your francium and the freezing point of francium?

____

____

30. Explain how adding 400 + 329 is different from adding 4.009 + 329. Then determine each sum.

____

____

Name: Date:
Instructor: Section:

Chapter 5 PROBLEM SOLVING WITH MIXED NUMBERS AND DECIMALS

Activity 5.6

Learning Objectives

1. Multiply and divide decimals.
2. Estimate products and quotients that involve decimals.

Practice Exercises

For #1-10, perform each calculation.

1. $-18.23 \cdot -4.8$

2. $132.4 \cdot -0.006$

1. ________

2. ________

3. $-25.74 \cdot 0.65$

4. $-89.51 \cdot -9.4$

3. ________

4. ________

5. $250.8 \cdot -0.012$

6. $-53.76 \cdot 0.24$

5. ________

6. ________

7. $-0.95 \cdot -2.04$

8. $0.007 \cdot 23.4$

7. ____________

8. ____________

9. $0.0105 \cdot -36.5$

10. $-0.34 \cdot -2.16$

9. ____________

10. ____________

For #11-20, perform each calculation.

11. $13.05 \div 4.5$

12. $31.13556 \div 87$

11. ____________

12. ____________

13. $16.905 \div 0.0003$

14. $1800 \div 0.06$

13. ____________

14. ____________

Name: Date:
Instructor: Section:

15. $0.00648 \div 0.018$

16. $204.255 \div 0.0034$

17. $-698.481 \div 0.006$

18. $-160.92 \div 0.45$

19. $24.544 \div 5.2$

20. $-25.328 \div 0.008$

21. Estimate the product of $(0.51)(312)$.

22. Find the exact product from Exercise #21.

15. ______________

16. ______________

17. ______________

18. ______________

19. ______________

20. ______________

21. ______________

22. ______________

23. Estimate the quotient $\frac{\$75,063}{0.47}$.

24. Find the quotient from Exercise #23, rounded to the nearest cent.

23. ______________

24. ______________

25. Estimate the quotient \$5461 divided by 0.62.

26. Find the quotient from Exercise #25, rounded to the nearest cent.

25. ______________

26. ______________

27. Estimate the product of 1492 and 3.14.

28. Find the exact product from Exercise #27.

27. ______________

28. ______________

Concept Connections

29. Bill multiplies 390.56 and 4.05 to get 158.1768. How can Bill use estimation to check his answer? Is his answer correct? If not, what is the correct answer?

____

____

30. Sarah divides 637.245 by 5.78 to get 1102.5. How can Sarah use estimation to check her answer? Is her answer correct? If not, what is the correct answer?

____

____

Name: Date:
Instructor: Section:

Chapter 5 PROBLEM SOLVING WITH MIXED NUMBERS AND DECIMALS

Activity 5.7

Learning Objectives
1. Use the order of operations to evaluate expressions that include decimals.
2. Use the distributive property in calculations that involve decimals.
3. Evaluate formulas that include decimals.
4. Solve equations of the form $ax = b$ and $ax + bx = c$ that involve decimals.

Practice Exercises

For #1-8, calculate using the order of operations.

1. $58.7 - 11.36 + 4.33$ **2.** $0.63 \div 0.09 \cdot 0.4$

1. ____________

2. ____________

3. $15.42 + 9.06 \cdot 0.5 - (0.06)^2$ **3.** ____________

4. $-3.8 \cdot 0.02 + (-0.03) \cdot (-9.5)$ **4.** ____________

5. $5.01 - 7 + 4.6 \cdot 0.8 - (0.003)^2$ **5.** ____________

6. $(-0.7)^2 - 4$ **6.** ____________

7. $0.3-0.55(6.8-7)^2$

8. $0.35 \div 0.07 \cdot 0.6$

7. ______________

8. ______________

For #9-12, evaluate each expression.

9. $\frac{1}{3}Bh$, where $B = 90.2$ and $h = 6.3$

10. $\frac{P-2l}{2}$, where $P = 67.4$, and $l = 18.8$

9. ______________

10. ______________

11. $\frac{V}{lw}$, where $V = 64.8$, $l = 3.6$, and $w = 2.4$

12. $0.05Pt$, where $P = 12{,}962$ and $t = 4$

11. ______________

12. ______________

For #13-17, translate each verbal statement into an equation and solve.

13. The quotient of a number and 7.2 is –5.1.

14. A number times 8.2 is 77.9.

13. ______________

14. ______________

Name: Date:
Instructor: Section:

15. 23.8 is the product of a number and –3.5.

16. A number divided by 3.2 is 16.4.

15. ______________

16. ______________

17. –8.3 times a number is 49.8

18. The quotient of a number and –4.8 is 2.9.

17. ______________

18. ______________

For #19-28, solve each equation. Round the result to the nearest tenth.

19. $5x = 17.5$

20. $-6.2x = 51.46$

19. ______________

20. ______________

21. $-7.3x = -8.76$

22. $17.6 = 17.6x$

21. ______________

22. ______________

23. $141.6 = 6x$

24. $5.3x = -\sqrt{81}$

23. ______________

24. ______________

25. $3.7x - 4.02x = -0.369$

26. $-0.7x + 3.2x = 95$

25. ______________

26. ______________

27. $5.1x + 3.61x = 8.71$

28. $-1.2x + 5.78x = -49$

27. ______________

28. ______________

Concept Connections

29. Explain the steps you would take to determine the value for P from the formula $A = P + Prt$ if $A = \$3357$, $r = 0.085$, and $t = 1.4$.

____

____

30. When John evaluates $4c \div 3b$ where $b = 0.4$ and $c = 1.2$, and incorrectly gets 4 as the answer. What did he do wrong? What is the correct answer?

____

____

Name: Date:
Instructor: Section:

Chapter 5 PROBLEM SOLVING WITH MIXED NUMBERS AND DECIMALS

Activity 5.8

Learning Objectives
1. Know the metric prefixes and their decimal values.
2. Convert measurements between metric quantities.

Key Terms
Use the vocabulary terms listed below to complete each statement in Exercises 1–3.

milligram **centigram** **kilogram**

1. A ____________________ is equal to 0.01 g.

2. A ____________________ is equal to 1000 g.

3. A ____________________ is equal to 0.001 g.

Practice Exercises
For #4-28, convert each amount.

4. Convert 6.7 meters into centimeters.

5. Convert 962 centimeters into meters.

6. Convert 3.8 meters into millimeters

7. Convert 7629 millimeters into meters.

8. Convert 5678 meters into kilometers.

9. Convert 8.73 kilometers into meters.

4. ________________

5. ________________

6. ________________

7. ________________

8. ________________

9. ________________

10. Convert 56.77 grams into centigrams.

11. Convert 843 grams into kilograms.

10. ________________

11. ________________

12. Convert 67.3 liters into milliliters.

13. Convert 982 milliliters into liters.

12. ________________

13. ________________

14. Convert 6.12 meters into centimeters.

15. Convert 6752 millimeters into meters.

14. ________________

15. ________________

16. Convert 6 centimeters into millimeters.

17. Convert 6572 meters into kilometers.

16. ________________

17. ________________

18. Convert 9877 milligrams into grams.

19. Convert 8.31 kilometers into meters.

18. ________________

19. ________________

20. Convert 675 milliliters into liters.

21. Convert 0.95 kilograms into grams.

20. ________________

21. ________________

Name: Date:
Instructor: Section:

22. Convert 8.8 kilometers into millimeters.

23. Convert 0.019 liters into milliliters.

22. ______________

23. ______________

24. Convert 18.2 milliliters into liters.

25. Convert 9.17 kilograms into grams.

24. ______________

25. ______________

26. Convert 3.88 kilometers into meters.

27. Convert 6.95 kilometers into millimeters.

26. ______________

27. ______________

28. Convert 0.038 liters into milliliters.

28. ______________

Concept Connections

29. State the basic metric units for measuring distance, mass and volume.

____

____

30. Explain why it is so easy to convert amounts in the metric system.

____

____

Name: Date:
Instructor: Section:

Chapter 6 PROBLEM SOLVING WITH RATIOS, PROPORTIONS AND DECIMALS

Activity 6.1

Learning Objectives

1. Understand the distinction between actual and relative measure.
2. Write a ratio in its verbal, fraction, decimal, and percent formats.

Key Terms

Use the vocabulary terms listed below to complete each statement in Exercises 1–6.

percent	**ratio**	**relative measure**
verbal	**fractional**	**decimal**

1. ____________________ is a word used to describe the relative measure quotient.

2. ____________________ is a quotient that compares two similar quantities, often a "part" and a "total." The part is divided by the total.

3. ____________________ always indicates a ratio out of 100.

4. A ratio can be expressed in the form ________________ $\left(\frac{4}{5}\right)$.

5. A ratio can be expressed in the form ________________ (4 out of 5).

6. A ratio can be expressed in the form ________________ (0.8).

Practice Exercises

For #7–12, write the ratio as a fraction.

7. 55 out of 103

8. 32 out of 99

7. ________________

8. ________________

9. 13 out of 20

10. 5 out of 9

9. ________________

10. ________________

11. 99 out of 100

12. 6 out of 7

11. ____________

12. ____________

For #13–18, write the fraction as a decimal.

13. $\frac{17}{20}$

14. $\frac{35}{80}$

13. ____________

14. ____________

15. $\frac{24}{96}$

16. $\frac{27}{70}$

15. ____________

16. ____________

17. $\frac{42}{90}$

18. $\frac{39}{52}$

17. ____________

18. ____________

For #19–23, write the ratio in percent format.

19. 37 out of 50

20. 9 out of 25

19. ____________

20. ____________

Name: Date:
Instructor: Section:

21. 7 out of 16 **22.** 9 out of 30

21. ______________

22. ______________

23. 9 out of 72

23. ______________

For #24–28, write the percent in decimal format.

24. 83% **25.** 4.5%

24. ______________

25. ______________

26. 350% **27.** 0.63%

26. ______________

27. ______________

28. 1.23%

28. ______________

Concept Connections

29. After a basketball game, Tyus said he made 12 baskets, and Jayson said he made 16 baskets. You want to determine the relative performance for each player. What information is missing?

____

____

30. From Exercise #29, suppose that Tyus had made 18 attempts, and Jayson had 22 attempts. Which player can claim a better performance?

____

____

Name: Date:
Instructor: Section:

Chapter 6 PROBLEM SOLVING WITH RATIOS, PROPORTIONS AND DECIMALS

Activity 6.2

Learning Objectives
1. Recognize that equivalent fractions lead to a proportion.
2. Use a proportion to solve a problem that involves ratios.

For #1-12, solve each proportion for x.

1. $\frac{3}{11}=\frac{x}{88}$

2. $\frac{9}{7}=\frac{54}{x}$

3. $\frac{x}{21}=\frac{100}{70}$

4. $\frac{5}{13}=\frac{x}{65}$

5. $\frac{7}{18}=\frac{21}{x}$

6. $\frac{153}{x}=\frac{17}{15}$

7. $\frac{x}{204}=\frac{87}{34}$

8. $\frac{78}{x}=\frac{138}{23}$

1. ______________

2. ______________

3. ______________

4. ______________

5. ______________

6. ______________

7. ______________

8. ______________

9. $\dfrac{120}{x} = \dfrac{3}{40}$

10. $\dfrac{360}{x} = \dfrac{9}{5}$

9. ______________

10. ______________

11. $\dfrac{13}{8} = \dfrac{x}{120}$

12. $\dfrac{16}{23} = \dfrac{48}{x}$

11. ______________

12. ______________

For #13-20, solve.

13. A car travels 102 miles on 3 gallons of gas. How far can it travel on 11 gallons of gas?

13. ______________

14. Three gallons of paint cover 1350 square feet of wall. How many gallons of paint are needed for 3150 square feet of wall?

14. 

15. Dan drove his new car 17,500 miles in the first 5 months. At this rate, how many miles will he drive in 2 years?

15. 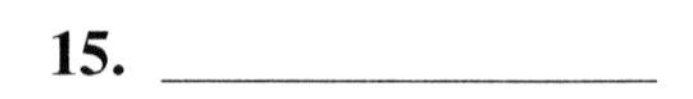

16. The scale for a map is 1.5 inches equals 50 miles. If two cities are 6 inches apart on the map, find the distance between them.

16. ______________

Name: Date:
Instructor: Section:

17. A stock split of 5 shares for every 2 shares owned. If Greg owned 350 shares, how many shares will he have after the split?

17. 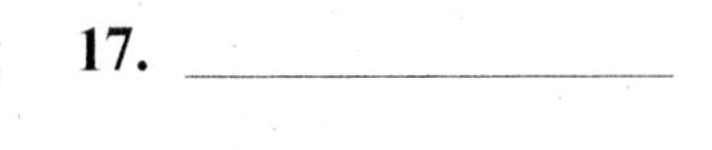

18. In a pizza dough recipe, 1.5 teaspoons of salt are mixed with 4 cups of flour. How much salt is needed for 20 cups of flour?

18. 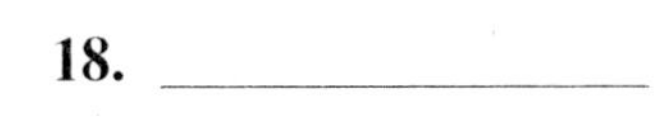

19. A Little League team won $\frac{3}{5}$ of the games played. They played 175 games. How many games did they win?

19. 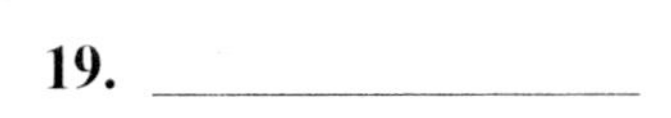

20. In an online math class 16 students earned an A, which was $\frac{4}{9}$ of the class. How many students were enrolled in the class?

20. 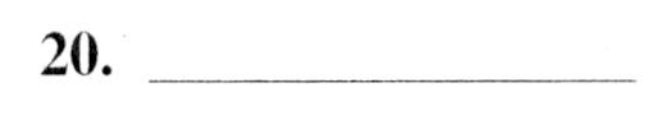

For #21-28, determine whether each proportion is true or false.

21. $\frac{27}{8} = \frac{18}{5}$

22. $\frac{225}{675} = \frac{3}{9}$

21. ______________

22. ______________

23. $\frac{39}{54} = \frac{52}{72}$

24. $\frac{7}{4} = \frac{42}{20}$

23. ________________

24. ________________

25. $\frac{450}{7} = \frac{300}{4}$

26. $\frac{28}{12} = \frac{7}{3}$

25. ________________

26. ________________

27. $\frac{304}{13} = \frac{47}{2}$

28. $\frac{57}{3} = \frac{133}{7}$

27. ________________

28. ________________

Concept Connections

29. Bill says that $\frac{a}{b} = \frac{c}{d}$ is equivalent to $\frac{d}{c} = \frac{b}{a}$. Matt disagrees. Who is right? Why?

____

____

30. Jody says that $\frac{a}{b} = \frac{c}{d}$ is equivalent to $\frac{a}{d} = \frac{b}{c}$. Barbara disagrees. Who is right? Why?

____

____

Name: Date:
Instructor: Section:

Chapter 6 PROBLEM SOLVING WITH RATIOS, PROPORTIONS AND DECIMALS

Activity 6.3

Learning Objectives

1. Use proportional reasoning to apply a known ratio to a given piece of information.
2. Write an equation using the relationship ratio · total = part and then solve the resulting equation.

Key Terms

Use the vocabulary terms listed below to complete each statement in Exercises 1–3.

proportional reasoning **total** **part**

1. The first step is to identify whether the given piece of information represents a part or a __________________ .

2. The second step is to substitute all of the known information into the formula ratio · total = __________________ to form in equation.

3. __________________ is the process by which you apply a known ratio to one piece of information to determine a related piece of information.

Practice Exercises

For 4–27, determine the value of each expression.

4. $15 \cdot \frac{4}{5}$

5. $24 \div \frac{3}{4}$

4. __________________

5. __________________

6. $\frac{1}{6}$ of 72

7. $\frac{2}{3}$ of 36

6. __________________

7. __________________

8. $32 \div \frac{4}{5}$

9. $72 \div \frac{8}{9}$

8. ____________________

9. ____________________

10. 20% of 70

11. 25% of 32

10. ____________________

11. ____________________

12. $120 \div 0.80$

13. $240 \div 0.30$

12. ____________________

13. ____________________

14. $35 \cdot \frac{4}{7}$

15. $48 \cdot \frac{5}{6}$

14. ____________________

15. ____________________

16. 35% of 840

17. 72% of 450

16. ____________________

17. ____________________

18. $960 \div 80\%$

19. $8550 \div 45\%$

18. ____________________

19. ____________________

Name: Date:
Instructor: Section:

20. $3000 \div \frac{8}{9}$

21. $1000 \div \frac{5}{7}$

22. 90% of 330

23. 23% of 420

24. $20{,}000 \div 40\%$

25. $3500 \div 20\%$

26. 19% of 650

27. $762 \div 25\%$

20. ______________

21. ______________

22. ______________

23. ______________

24. ______________

25. ______________

26. ______________

27. ______________

Concept Connections

28. There are 480 children at a local elementary school. If 15% of the children are left-handed, find the number of children that are left-handed.

---

29. From Exercise, #28, assuming that the children at the local elementary school are either left-handed or right-handed, how many children are right-handed?

---

30. From Exercise #29, what percent of children at the local elementary school are right-handed?

Name: Date:
Instructor: Section:

Chapter 6 PROBLEM SOLVING WITH RATIOS, PROPORTIONS AND DECIMALS

Activity 6.4

Learning Objectives
1. Define actual change and relative change.
2. Distinguish between actual and relative change.

Key Terms
Use the vocabulary terms listed below to complete the equation for Exercises 1–2.

actual change **relative change**

(1)________________ = (2)________________ / original value

Practice Exercises
For 3–27, solve.

3. Last year, the Lab fee for a course was $50. This year the Lab fee for the same course is $65. Determine the actual increase in the Lab fee. **3.** ____________

4. From Exercise #3, determine the relative increase in the Lab fee. **4.** ____________

5. Last year, the textbook for a statistics class was $144. This year the textbook for the same class is $168. Determine the actual increase in price. **5.** ____________

6. From Exercise #5, determine the relative increase in price. **6.** ____________

7. Last year, the textbook for a history class was $95. This year the textbook for the same class is $115. Determine the actual increase in price.

7. ________________

8. From Exercise #7, determine the relative increase in price.

8. ________________

9. From Exercises #5 and #7, which textbook had the larger percentage increase?

9. ________________

10. The fall semester had an enrollment of 5200 students. The spring semester there were only 5035 students. Determine the actual decrease in enrollment.

10. ________________

11. From Exercise #10, determine the relative decrease (as a percent) in enrollment.

11. ________________

Name: Date:
Instructor: Section:

12. The first summer session had an enrollment of 1520. The second summer session had only 925 students. Determine the actual decrease in enrollment.

12. ________________

13. From Exercise #12, determine the relative decrease (as a percent) in enrollment.

13. ________________

14. From Exercises #10 and #12, which summer session had the largest decrease in enrollment?

14. ________________

15. Before Tyler joined a gym, he weighed 247 lbs. After a year in the exercise program, he weighed 212 lbs. Determine the actual decrease in weight.

15. ________________

16. From Exercise #15, determine the relative decrease (as a percent) in weight.

16. ________________

17. Tyler quit the gym when he weighed 212 lbs and gained weight. He now weighs 247 lbs. Determine the increase in weight.

17. ________________

18. From Exercise #17, determine the relative increase (as a percent) in weight.

18. ________________

19. A prized rose bush has an average bloom of 15. Using a new fertilizer the number of blooms increases to 21. Determine the actual increase in blooms.

19. ________________

20. From Exercise #19, determine the relative increase (as a percent) in blooms.

20. ________________

21. A prized fertilized rose bush has an average bloom of 21. Lack of water decreases the number of blooms to 8. Determine the actual decrease in blooms.

21. ________________

22. From Exercise #21, determine the relative decrease (as a percent) in blooms.

22. ________________

Name: Date:
Instructor: Section:

23. From Exercises #19 and #21, does fertilizer or lack of water cause the largest change in blooms?

23. ______________

24. The average price of a gallon of gasoline was \$2.77. During the vacation season a gallon of gasoline was \$2.95. Determine the actual increase in the price of gasoline.

24. ______________

25. From Exercise #24, determine the relative increase (as a percent) in the price.

25. ______________

26. After the vacation season, the price of a gallon of gasoline drops to \$2.77 from \$2.95. Determine the actual decrease in the price of gas.

26. ______________

27. From Exercise #26, determine the relative decrease (as a percent) in the price.

27. ______________

Concept Connections

28. You buy stock with a company for $50. The stock price increases to $150. Find the percent increase. Why is this percent more than 100%?

29. You buy stock with a company for $100. The stock price decreases to $25. Find the percent decrease. When dealing with the stock market, will a percent decrease ever be less than 0%?

30. You buy stock with a company for $58. Under what circumstances would the percent decrease reach 100%?

Name: Date:
Instructor: Section:

Chapter 6 PROBLEM SOLVING WITH RATIOS, PROPORTIONS AND DECIMALS

Activity 6.5

Learning Objectives
1. Define and determine growth factors.
2. Use growth factors in problems that involve percent increases.

Key Terms

Use the vocabulary terms listed below to complete each statement in Exercises 1–3.

growth factor **multiplying** **dividing**

1. When a quantity increases by a specified percent, its new value can be obtained by ____________________ the original value by the corresponding growth factor.

2. When a quantity has already increased by a specified percent, its original value is obtained by ____________________ the new value by the corresponding growth factor.

3. When a quantity increases by a specified percent, the ratio of its new value to the original value is called the ____________________ .

Practice Exercises

For #4-16, determine the growth factor of a quantity that increases by the given percent. Write the result as a percent and then as a decimal.

4. 85% **5.** 40%

4. ____________________

5. ____________________

6. 18% **7.** 3%

6. ____________________

7. ____________________

8. 2.6% **9.** 222%

8. ____________________

9. ____________________

10. 99% **11.** 0.2%

10. ________________

11. ________________

12. 2000% **13.** 0.5%

12. ________________

13. ________________

14. 21.7% **15.** 93%

14. ________________

15. ________________

16. 600%

16. ________________

For 17–28, solve.

17. You are purchasing a new car. The purchase price of the car in Sacramento, California is $21,995 excluding sales tax. The sales tax rate is 8.75%. What growth factor is associated with the sales tax rate?

17. ________________

18. From Exercise #17, use the growth factor to determine the total cost of the car.

18. ________________

19. The cost of bus fare in a small city is now $2. Last June the price was increased by 14.3%. What was the pre-June price of the fare? Round to the nearest penny.

19. ________________

Name: Date:
Instructor: Section:

20. One year, a city high school had enrollment of 1195 students. The next year enrollment increased to 1244. What is the ratio of the enrollment in the second year to the enrollment in the first year?

20. ________________

21. From Exercise #20, determine the growth factor. Write in decimal form to the nearest thousandth.

21. ________________

22. From Exercise #20, by what percent did the enrollment increase?

22. ________________

23. The number of registered voters in a city in Oregon increased by 13.5% from 2000 to 2008. In 2008, the number of registered voters was reported to be 27,919. What was the number of registered voters in the Oregon city in 2000?

23. ________________

24. In 1979, your grandparents bought a house for $50,000. Due to inflation, the cost of the same house in 2009 was $157,500. What is the inflation growth factor of housing from 1979 to 2009?

24. ________________

25. From Exercise #24, what is the inflation rate of housing from 1979 to 2009?

25. ________________

26. From Exercise #24, the house was sold in 2009 for $189,000. What profit was made in terms of 2009 dollars?

26. ________________

27. Your friend plans to move from Alabama to Connecticut. He earns $40,000 a year in Alabama. How much must he earn in Connecticut to maintain the same standard of living if the cost of living in Connecticut is 38% higher?

27. ________________

28. A one-year certificate of deposit earns an annual percentage yield of 4.05%. If $4000 is invested, how much will the investment be worth in a year?

28. ________________

Concept Connections

29. After applying a growth factor, which is larger, the original value or the new value? Why?

____

____

30. In what three forms can a growth factor be represented?

____

____

Name: Date:
Instructor: Section:

Chapter 6 PROBLEM SOLVING WITH RATIOS, PROPORTIONS AND DECIMALS

Activity 6.6

Learning Objectives
1. Define and determine decay factors.
2. Use decay factors in problems that involve percent decreases.

Key Terms
Use the vocabulary terms listed below to complete each statement in Exercises 1–3.

decay factor **subtracting** **multiplying** **dividing**

1. When a quantity has already decreased by a specified percent, its original value is obtained by ____________________ the new value by the corresponding decay factor.

2. When a quantity decreases by a specified percent, its new value can be obtained by ____________________ the original value by the corresponding decay factor.

3. A ____________________ is determined by ____________________ the specified decrease from 100% and changing the resulting percent into decimal form.

Practice Exercises
For #4-16, determine the decay factor of a quantity that decreases by the given percent. Write the result as a percent and then as a decimal.

4. 85% **5.** 40%

4. ________________

5. ________________

6. 18% **7.** 3%

6. ________________

7. ________________

8. 2.6% **9.** 22%

8. ________________

9. ________________

10. 0.2% **11.** 99%

10. ______________

11. ______________

12. 98% **13.** 0.5%

12. ______________

13. ______________

14. 21.7% **15.** 93%

14. ______________

15. ______________

16. 1.1%

16. ______________

For 17–28, solve.

17. In 1998 in a city in Ohio, the number of children enrolled in the city preschool program was 20,162. In 2008, the number of children enrolled in the city preschool program dropped to 19,073. For the 10-year period, what is the decay factor for the city preschool program?

17. ______________

18. From Exercise #17, what is the percent decrease?

18. ______________

Name: Date:
Instructor: Section:

19. You wrote a 24-page booklet to be published. The editor asked you to reduce the number of pages to 20. By what percent must you reduce the length of the booklet?

19. ______________

20. A deluxe paper shredder has a list price of $149.99. On Sunday, the model goes on sale and the price is reduced by 20%. What is the decay factor?

20. ______________

21. From Exercise #20, use the decay factor to determine the sale price of the deluxe paper shredder.

21. ______________

22. You are on a weight-reducing diet. Your goal is to lose 15% of your body weight over the next year. You currently weigh 235 pounds. What is the decay factor?

22. ______________

23. From Exercise #22, use the decay factor to calculate the goal weight to the nearest pound.

23. ______________

24. In general, 75% of a dose of aspirin is eliminated from the blood stream in an hour. A person takes a low dose of aspirin of 81 milligrams. How much aspirin remains in the person after 1 hour?

24. ______________

25. You are flying from Washington D.C. to Chicago. The restricted, nonrefundable fare is $176.40. You book online to take advantage of the 5% discount on the ticket. What is the decay factor?

25. ______________

26. From Exercise #25, determine the cost of the fare if you book online.

26. ______________

27. A deluxe wireless router has a list price of $69.99. Yesterday, the model went on sale and the price was reduced by 25%. What is the decay factor?

27. ______________

28. From Exercise #27, use the decay factor to determine the sale price of the deluxe wireless router.

28. ______________

Concept Connections

29. After applying a decay factor, which is larger, the original value or the new value? Why?

____

____

30. Give the definition of a decay factor.

____

____

Name: Date:
Instructor: Section:

Chapter 6 PROBLEM SOLVING WITH RATIOS, PROPORTIONS AND DECIMALS

Activity 6.7

Learning Objectives

1. Apply consecutive growth and/or decay factors to problems that involve two or more percent changes.

Key Terms

Choose the correct term listed in parentheses to complete each statement in Exercises 1–2.

1. The cumulative effect of a sequence of percent changes is the ____________________ (product/quotient) of the associated growth or decay factors.

2. With the cumulative effect of a sequence, the order in which the changes are applied ____________________ (does/does not) matter.

Practice Exercises

For 3–28, solve.

3. A $150 leather purse is on sale for 25% off. Determine the decay factor for the sale. **3.** ________________

4. From Exercise #3, you present a coupon for an additional 15% off. Determine the decay factor for the additional discount. **4.** ________________

5. From Exercises #3 and #4, use these decay factors to determine the price you pay for the purse. **5.** ________________

6. A pair of jeans has a regular price of $79.99. The sale is for 35% off. Determine the decay factor for the sale.

6. ________________

7. From Exercise #6, you present a coupon for an additional 10% off. Determine the decay factor for the additional discount.

7. ________________

8. From Exercises #6 and #7, use the decay factors to determine the price you pay for the jeans.

8. ________________

9. Your union has negotiated a 2-years contract containing annual raises of 3.5% and 4.5% during the term of the contract. Your current salary is $52,500. What salary will you earn in 2 years?

9. ________________

10. Your union has negotiated a 3-year contract containing annual raises of 2.5%, 3%, and 3.5% during the term of the contract. Your current salary is $49,000. What salary will you earn in 3 years?

10. ________________

Name: Date:
Instructor: Section:

11. As a vendor at an open air market, you increased your inventory of veggie yogurt cheese of 115 pounds by 15%. You sold 85% of your inventory. How many pounds of cheese remain?

11. ______________

12. As the owner of a news stand, you anticipate a large demand for a popular sports magazine and increase your inventory of 250 by 20%. You sold 70% of your inventory. How many sports magazines remain?

12. ______________

13. You deposit $2500 in a 3-year certificate of deposit that pays 3.5% interest compounded annually. Determine, to the nearest dollar, your account balance when your certificate comes due.

13. ______________

14. You deposit $3500 in a 4-year certificate of deposit that pays 3% interest compounded annually. Determine, to the nearest dollar, your account balance when your certificate comes due.

14. ______________

15. Determine a single decay factor that represents the cumulative effect of consecutively applying discounts of 50% and 45%.

15. ________________

16. From Exercise #15, use the decay factor to determine the effective discount.

16. ________________

17. Determine a single decay factor that represents the cumulative effect of consecutively applying discounts of 30%, 40%, and 50%.

17. ________________

18. From Exercise #17, use this decay factor to determine the effective discount.

18. ________________

19. A charitable organization has decreased its budget of $450,000 by 6% in each of the last 3 years. What is the current budget?

19. ________________

Name: Date:
Instructor: Section:

20. A small church has decreased its budget of $59,000 by 4.5% in each of the last 4 years. What is the current budget?

20. ________________

21. An appliance store has decreased its budget of $117,600 for advertising by 8% in each of the last 4 years. What is the current budget?

21. ________________

22. A high growth stock is purchased for $2500. The value rose by 25%. Determine the growth factor.

22. ________________

23. After a year, the value of a high growth stock from Exercise #22 drops by 25%. Determine the decay factor.

23. ________________

24. From Exercises #22 and #23, what is the single factor that represents the increase and decrease?

24. ________________

25. From Exercise #24, does the number represent a growth factor or a decay factor? Why?

25. ________________

____

____

26. From Exercises #22 and #23, what is the current value of the high growth stock?

26. ________________

27. From Exercises #22 – 26, what is the cumulative effect of applying a 25% increase followed by a 25% decrease?

27. ________________

28. From Exercises #22 – 26, what is the cumulative effect if the 25% decrease had been applied first, followed by the 25% increase?

28. ________________

____

____

Concept Connections

29. Which is a better deal: 15% discount and additional 40% discount, or 20% discount and additional 30% discount?

____

____

30. Which is a better deal: 30% discount or 10% discount and additional 20% discount?

____

____

Name: Date:
Instructor: Section:

Chapter 6 PROBLEM SOLVING WITH RATIOS, PROPORTIONS AND DECIMALS

Activity 6.8

Learning Objectives

1. Apply rates directly to solve problems.
2. Use proportions to solve problems that involve rates.
3. Use unit analysis or dimensional analysis to solve problems that involve consecutive rates.

For 1–28, convert.

1. Convert 15 hours to seconds. **1.** ______________

2. Convert 2 meters to inches. **2.** ______________

3. Convert 4 km to miles. **3.** ______________

4. Convert 80 km/hr to mph. **4.** ______________

5. Convert 180 cm to inches. **5.** ______________

6. Convert 33,840 seconds to hours. **6.** ______________

7. Convert 2.5 weeks to minutes. **7.** ______________

8. Convert 3L/day to milliliters/hour. **8.** ______________

9. Convert 48 mg/day to grams in 30 days. **9.** ______________

10. Convert 2.4L/day to milliliters per hour. **10.** ______________

11. Convert 162 inches to yards. **11.** ______________

Name: Date:
Instructor: Section:

12. Convert 2640 feet to miles.

12. ______________

13. Convert 3 miles to inches.

13. ______________

14. Convert 1.5 yards to meters

14. ______________

15. Convert 35 mg/week to grams/12 weeks

15. ______________

16. Convert 5 weeks to minutes.

16. ______________

17. A marathon race is 26.2 miles. Find the distance in feet. **17.** ________________

18. What is the distance of the marathon in kilometers? **18.** ________________

19. A person who weighs 120 pounds on Earth would weigh 19.8 pounds on the moon. What would be a person's weigh on the moon if she weighs 140 pounds on Earth? **19.** ________________

20. Your car's highway fuel efficiency is 34 miles per gallon. What is its fuel efficiency in kilometers per liter? **20.** ________________

21. Your SUV's highway fuel efficiency is 25 miles per gallon. What is its fuel efficiency in kilometers per liter? **21.** ________________

22. Your recreational vehicle's highway fuel efficiency is 12 miles per gallon. What is its efficiency in kilometers per liter? **22.** ________________

Name: Date:
Instructor: Section:

23. A soda bottle contains 2 liters of cola. How many quarts of cola are in the soda bottle?

23. ________________

24. A 2-liter bottle of cola sells for $1.58. What is the price for a 3-liter bottle of cola?

24. ________________

25. A sprinter runs the 200 yard dash. What is the length in feet?

25. ________________

26. How many seconds are there in 3 days?

26. ________________

27. How many hours are in 3 weeks?

27. ________________

28. Your lawnmower uses a gasoline and oil mixture for fuel. For every 2 gallons of gasoline you add 18 ounces of oil. How much oil should you add for 5 gallons of gasoline?

28. ________________

Concept Connections

29. In the movie Julie and Julia, Julia states that when writing the American cookbook, one difficulty she had with replicating French recipes, was getting the amounts correct. What was she talking about?

____

____

30. At the grocery store, many packaged products contain more than one unit of measure. For example, a leading brand of peanut butter claims to contain 18 oz, or 510 g. How can you use this information to determine a rate of unit analysis to convert between ounces and grams?

____

____

Name: Date:
Instructor: Section:

Chapter 7 PROBLEM SOLVING WITH GEOMETRY

Activity 7.1

Learning Objectives

1. Recognize perimeter as a geometric property of plane figures.
2. Write formulas for, and calculate perimeters of, squares, rectangles, triangles, parallelograms, trapezoids, and polygons.
3. Use unit analysis to solve problems that involve perimeters.

Key Terms

Use the vocabulary terms listed below to complete each statement in Exercises 1–4.

rectangle **square** **trapezoid** **triangle**

1. The general formula for the perimeter of a ________________ is $P = a + b + c + d$.

2. The general formula for the perimeter of a ________________ is $P = 4s$.

3. The general formula for the perimeter of a ________________ is $P = 2l + 2w$.

4. The general formula for the perimeter of a ________________ is $P = a + b + c$.

Practice Exercises

For #5-6, find the perimeter of each polygon.

5.

30 ft

80 ft

6.

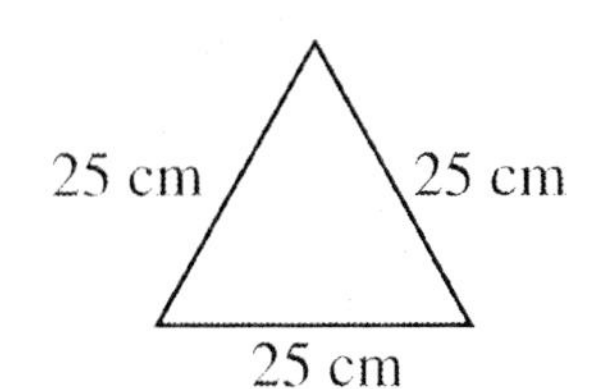

5. ________________

6. ________________

7. Identify the polygon from Exercise #5.

8. Identify the polygon from Exercise #6.

7. ________________

8. ________________

For #9-10, find the perimeter of each polygon.

9.

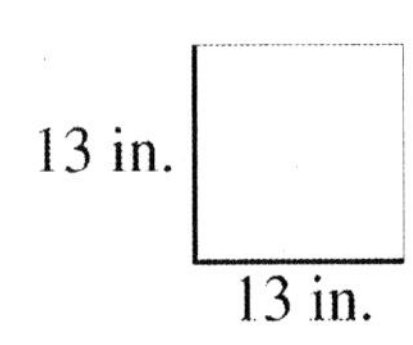

10.

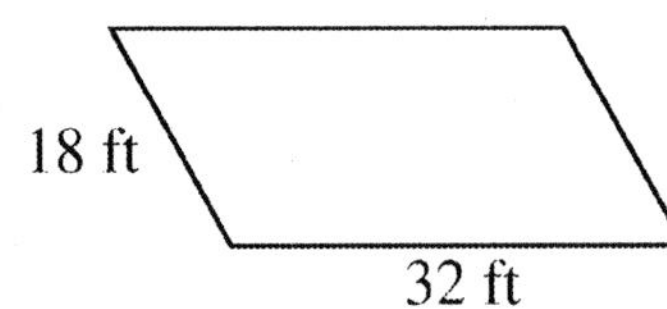

9. ________________

10. ________________

11. Identify the polygon from Exercise #9.

12. Identify the polygon from Exercise #10.

11. ________________

12. ________________

For #13-14, find the perimeter of each polygon.

13.

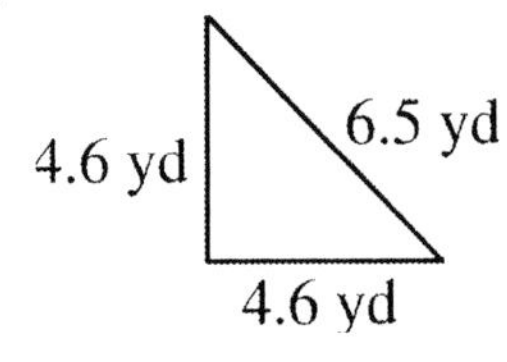

14.

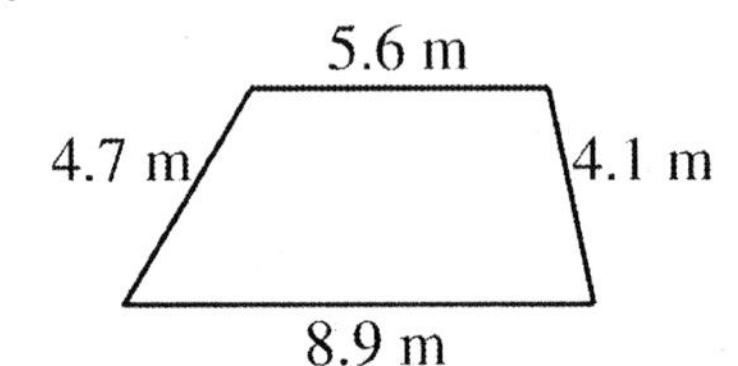

13. ________________

14. ________________

15. Identify the polygon from Exercise #13.

16. Identify the polygon from Exercise #14.

15. ________________

16. ________________

Name: Date:
Instructor: Section:

For #17-24, find the perimeter of each polygon.

17.

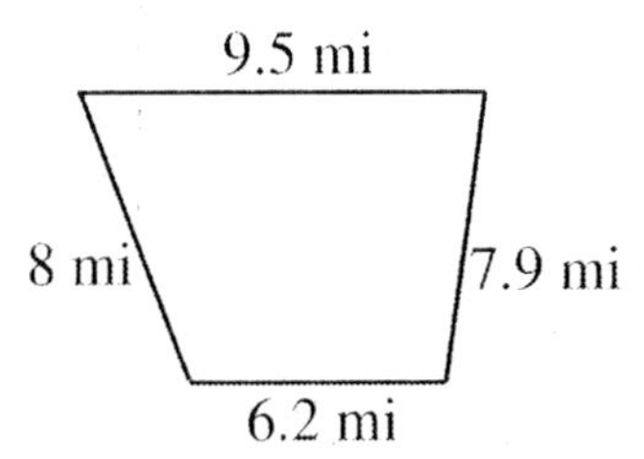

18.

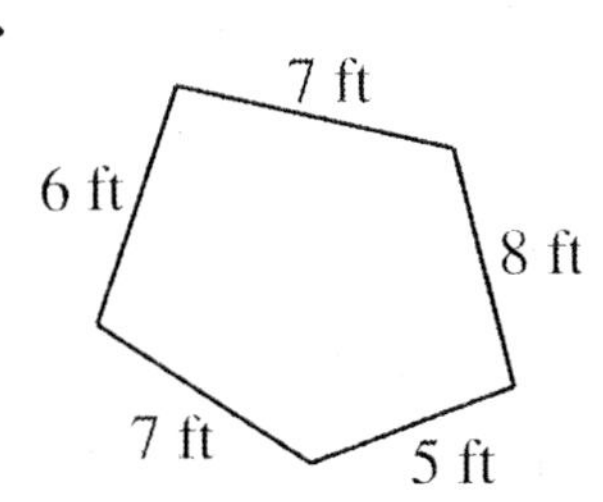

17. ______________

18. ______________

19.

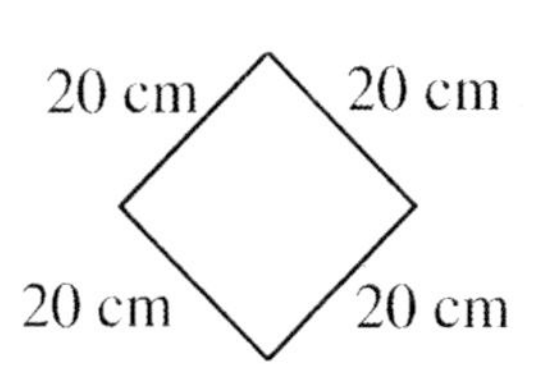

20.

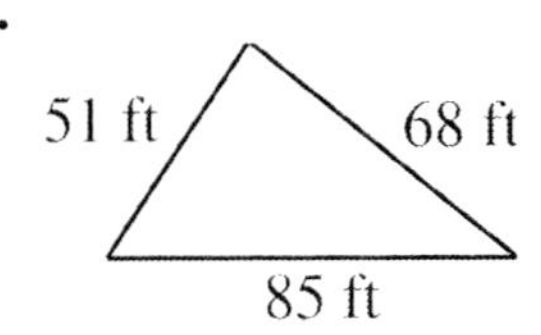

19. ______________

20. ______________

21.

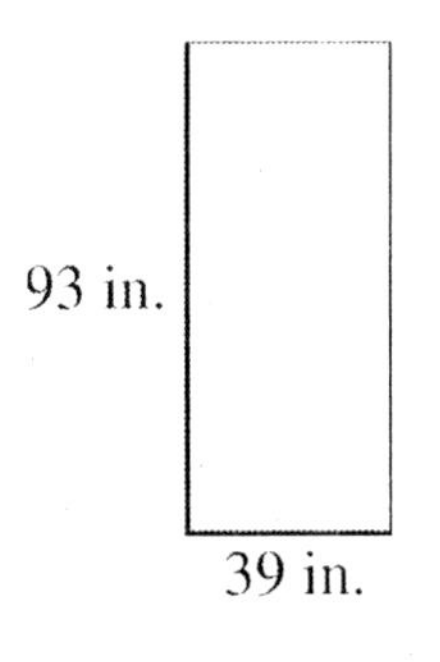

22.

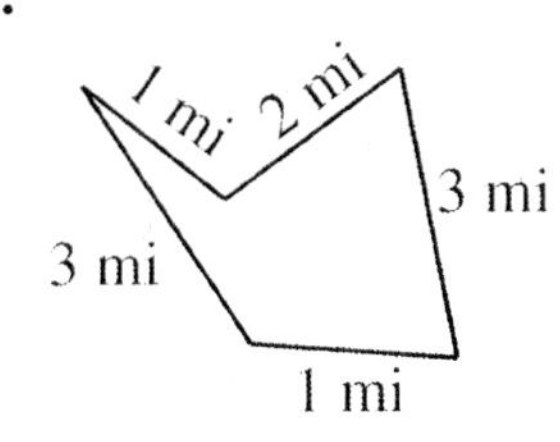

21. ______________

22. ______________

23.

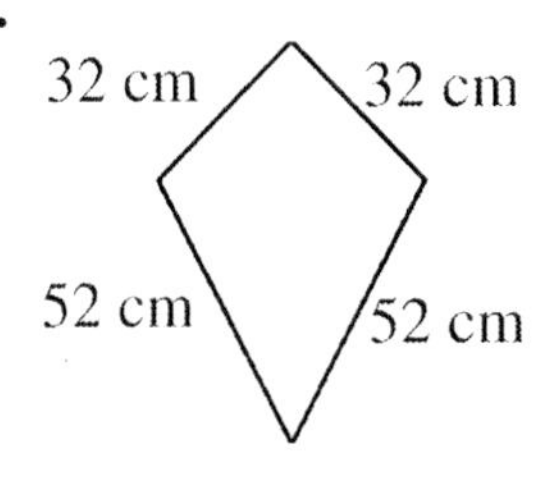

24.

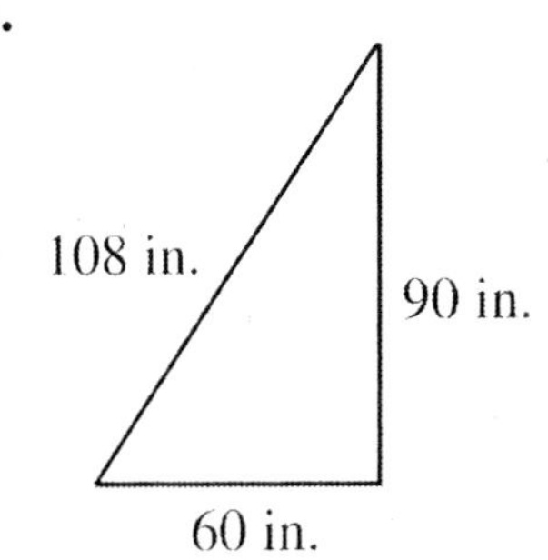

23. ______________

24. ______________

25. A painting has a perimeter of 72 ft. One side has a width of 12 ft. Calculate the length of the painting.

25. ______________

26. A scalene triangle has perimeter 124 inches. One side has a length of 31 inches, and another side has a length of 52 inches. Calculate the length of the third side.

26. ______________

27. A rectangular room has perimeter 68 ft. One side has a length of 14 ft. Calculate the width of the room.

27. ______________

28. A square table has perimeter 22 ft. Calculate the length of each side of the square.

28. ______________

Concept Connections

29. What is the difference between a trapezoid and a parallelogram?

____

____

30. What is the difference between an equilateral triangle and a scalene triangle?

____

____

Name: Date:
Instructor: Section:

Chapter 7 PROBLEM SOLVING WITH GEOMETRY

Activity 7.2

Learning Objectives
1. Develop and use formulas for calculating circumferences of circles.

Key Terms

Use the vocabulary terms listed below to complete each statement in Exercises 1–3.

circumference **diameter** **radius**

1. The ____________________ of a circle is a line segment that starts at its center and ends on its edge.

2. The distance around the edge of a circle is called the ____________________ .

3. The line segment that starts on one edge of a circle, passes through the center and ends on another edge of the circle is called the ____________________ .

Practice Exercises

For #4-9, find the diameter for each circle, given the radius.

4. radius = 12.5 in. **5.** radius = 8.3 miles

6. radius = 30 cm **7.** radius = 4.61 m

8. radius = 9 ft **9.** radius = 7.9 mm

4. ________________

5. ________________

6. ________________

7. ________________

8. ________________

9. ________________

For #10-15, find the radius for each circle, given the diameter.

10. diameter = 62.3 m

11. diameter = 18 ft

12. diameter = 40.8 mm

13. diameter = 29 in.

14. diameter = 0.74 mi

15. diameter = 30 cm

10. ______________

11. ______________

12. ______________

13. ______________

14. ______________

15. ______________

For #16-21, find the circumference for each circle given the radius or diameter. Round to the nearest tenth.

16. radius = 8.3 miles

17. diameter = 40.8 mm

18. diameter = 30 cm

19. radius = 9 ft

20. radius = 12.5 in.

21. diameter = 63.2 m

16. ______________

17. ______________

18. ______________

19. ______________

20. ______________

21. ______________

Name: Date:
Instructor: Section:

For #22-27, find the circumference of each circle. Round to the nearest tenth.

22.

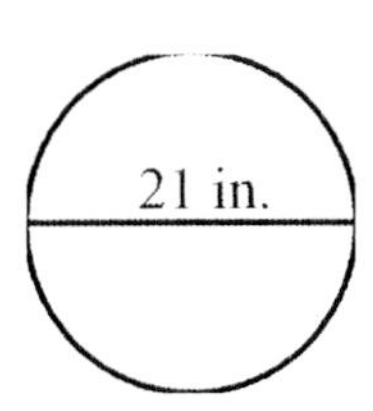

23.

7 mi

22. ____________

23. ____________

24.

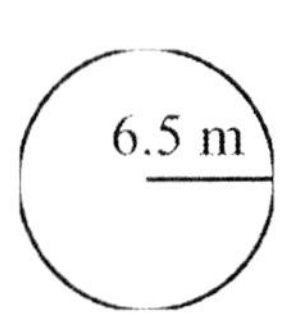

25.

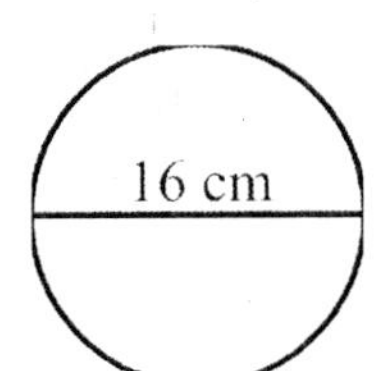

24. ____________

25. ____________

26.

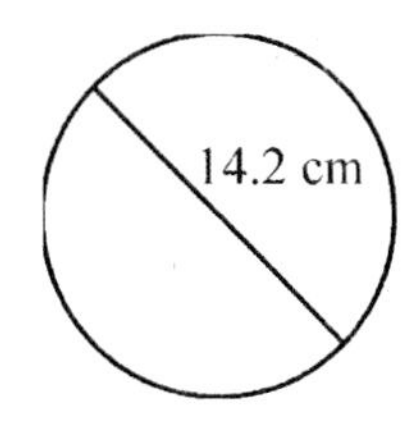

27.

26. ____________

27. ____________

28. A circle has circumference 84 in. Find the radius. Round to the nearest tenth.

28. ____________

Concept Connections

29. What is the formula for the perimeter of a circle?

30. What is the ratio of the circumference of any circle to its diameter?

Name: Date:
Instructor: Section:

Chapter 7 PROBLEM SOLVING WITH GEOMETRY

Activity 7.3

Learning Objectives

1. Calculate perimeters of many-sided plane figures using formulas and combinations of formulas.
2. Use unit analysis to solve problems that involve perimeters.

Practice Exercises

For #1-6, calculate the perimeters of each of the following figures.
Note: all sides of each regular figure have equal length.

1.

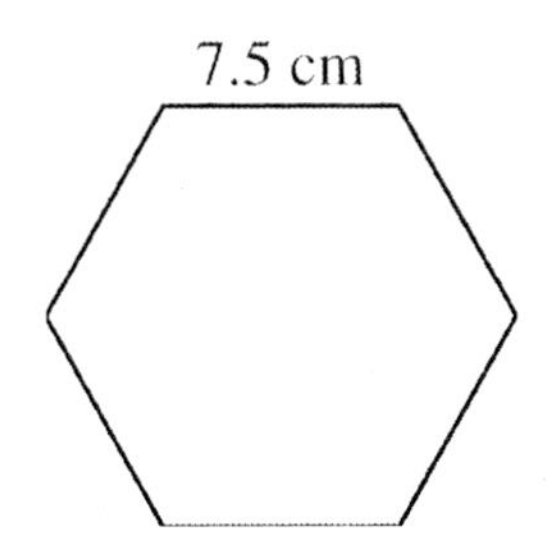

2.

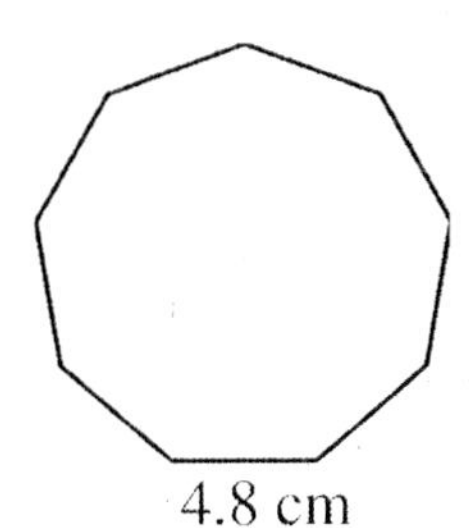

1. ______________

2. ______________

3.

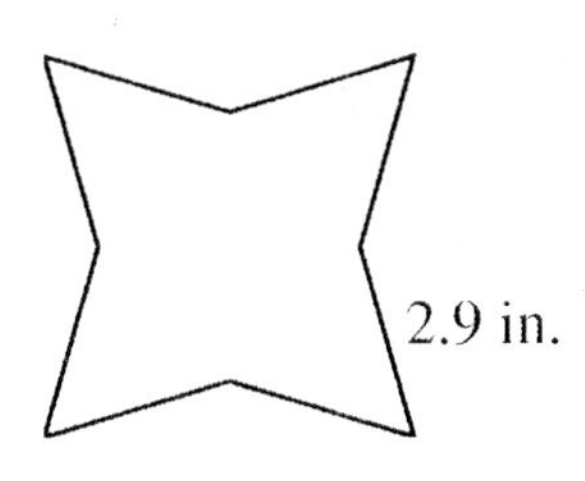

4.

3. ______________

4. ______________

5.

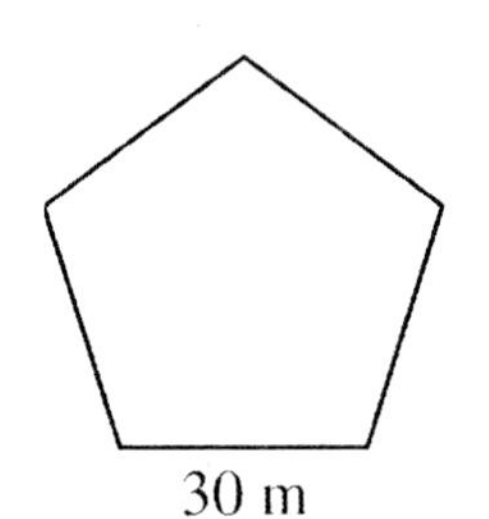

6.

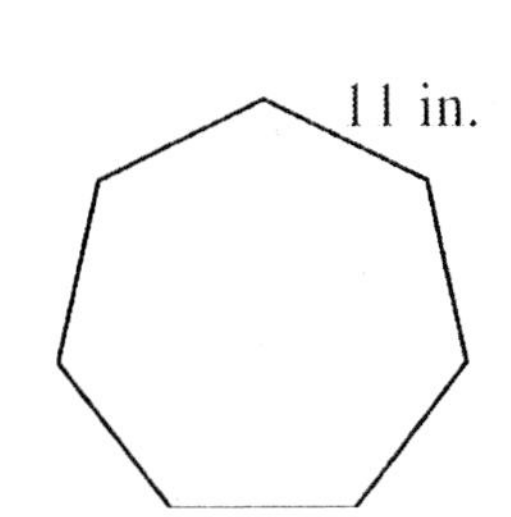

5. ______________

6. ______________

For #7-25, calculate the perimeters of each of the following figures. Round to the nearest tenth.

7.

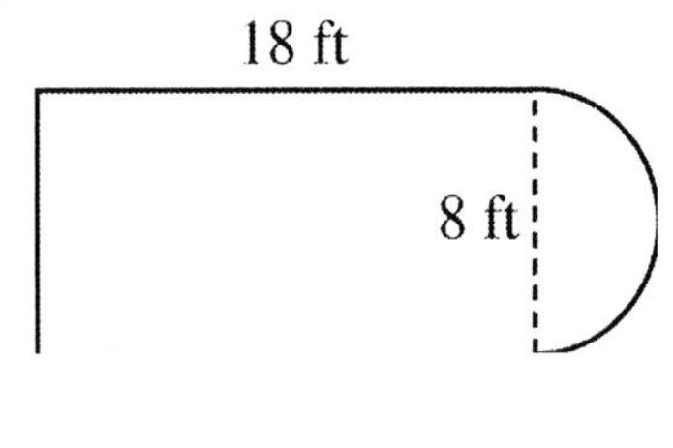

8.

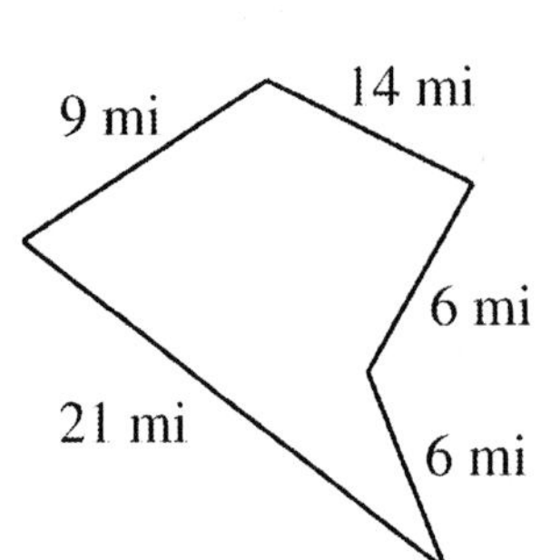

7. ____________

8. ____________

9.

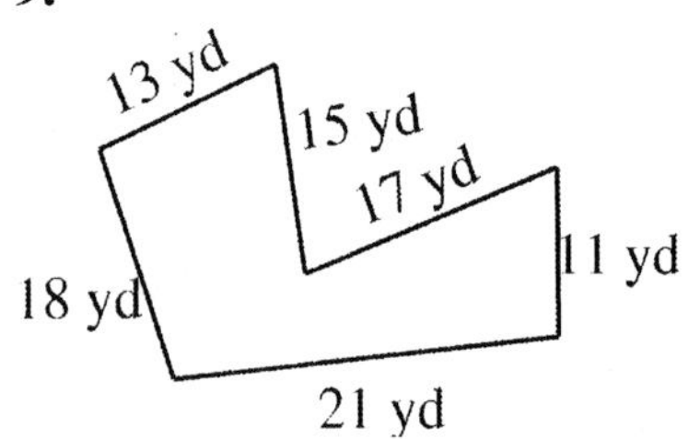

10.

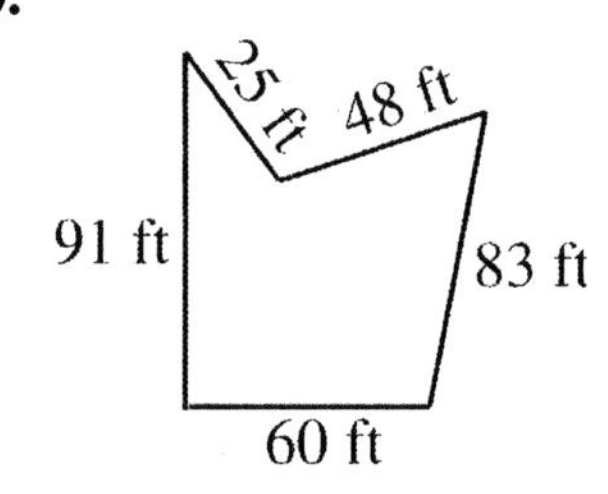

9. ____________

10. ____________

11.

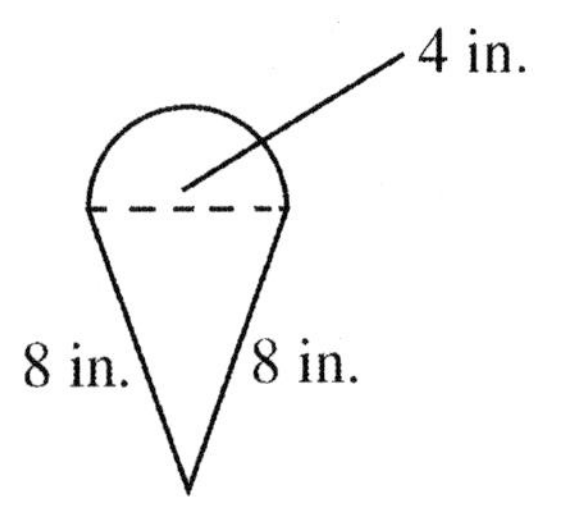

12.

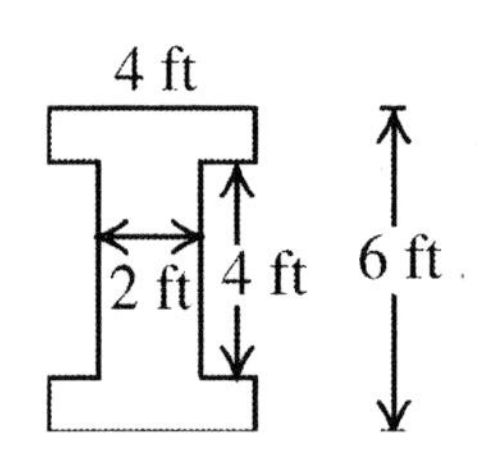

11. ____________

12. ____________

13.

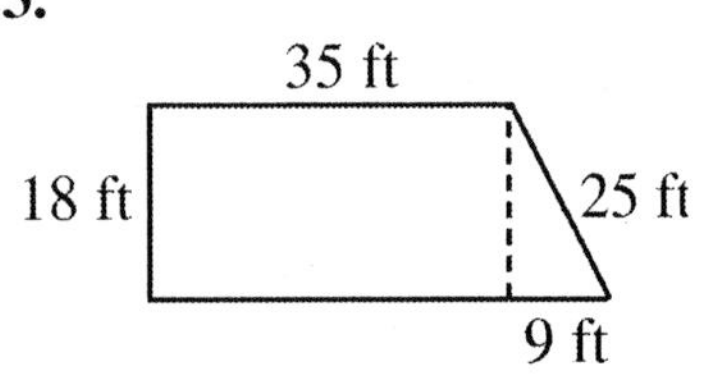

14.

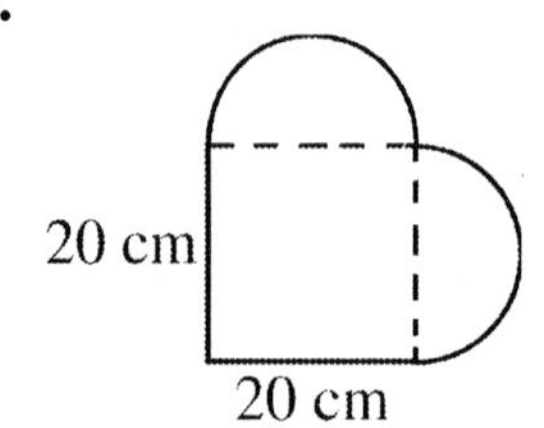

13. ____________

14. ____________

Name: Date:
Instructor: Section:

15.

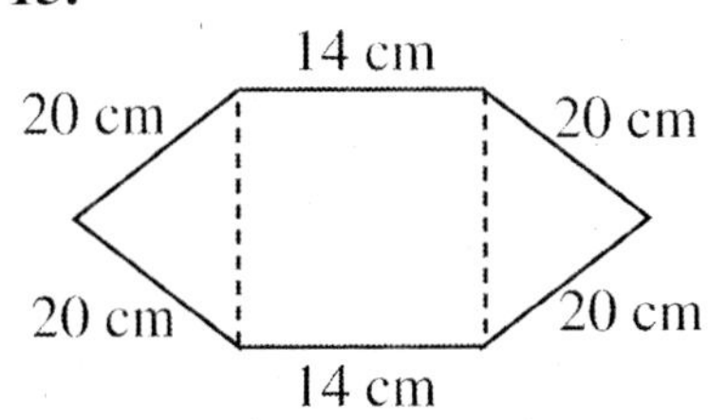

16.

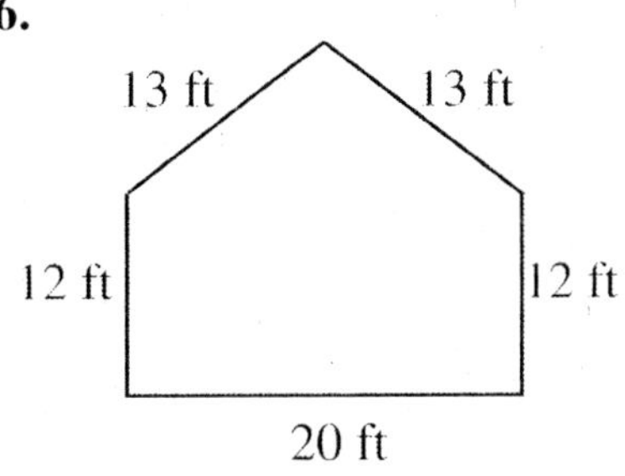

15. ____________

16. ____________

17.

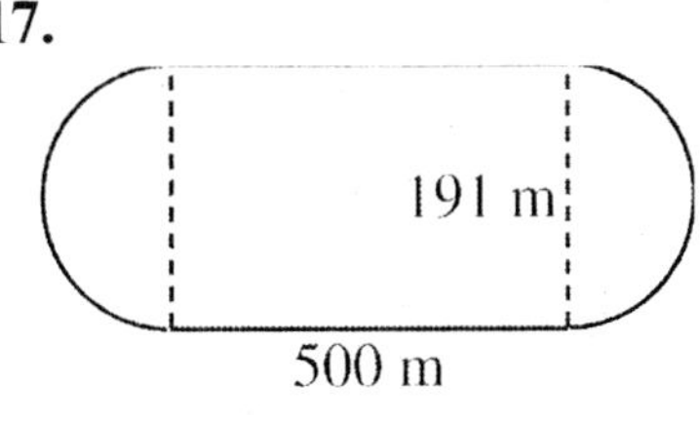

18.

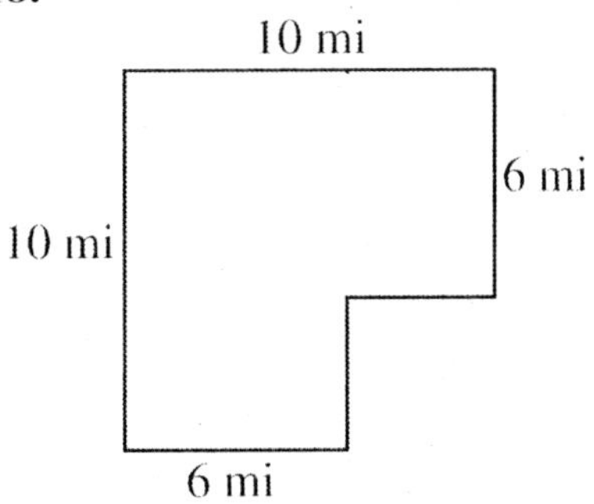

17. ____________

18. ____________

19.

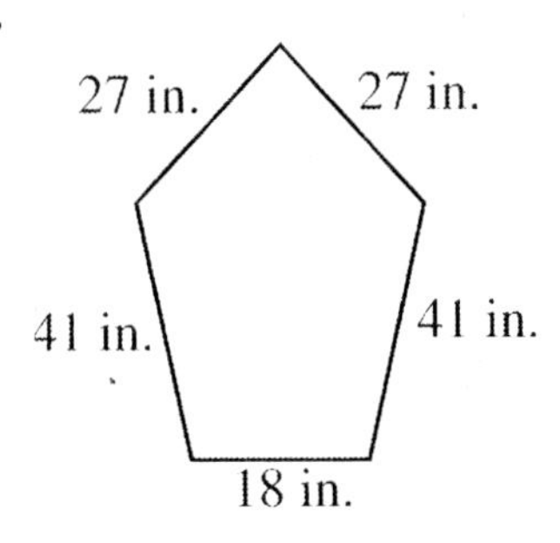

20.

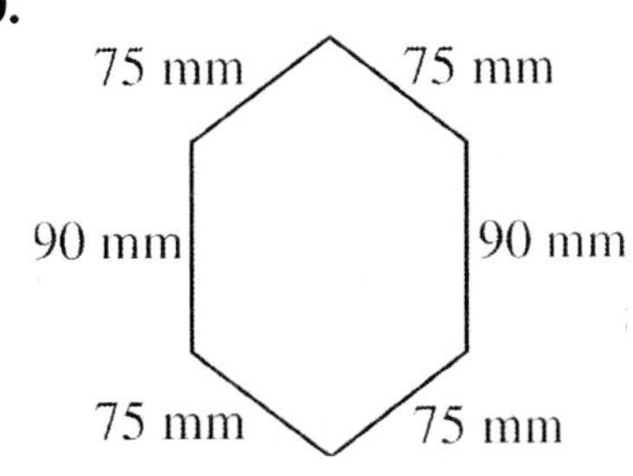

19. ____________

20. ____________

21.

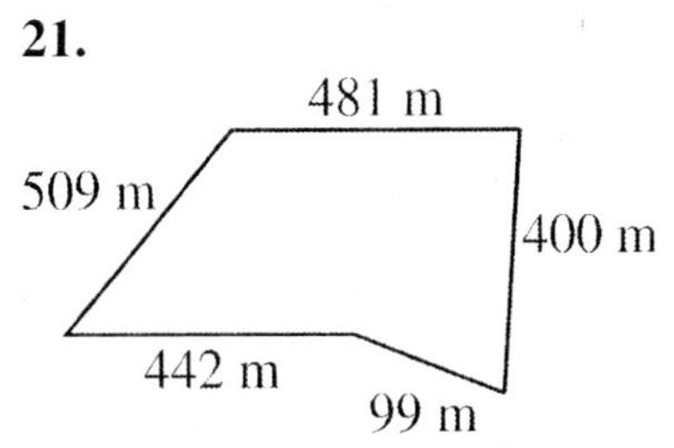

22.

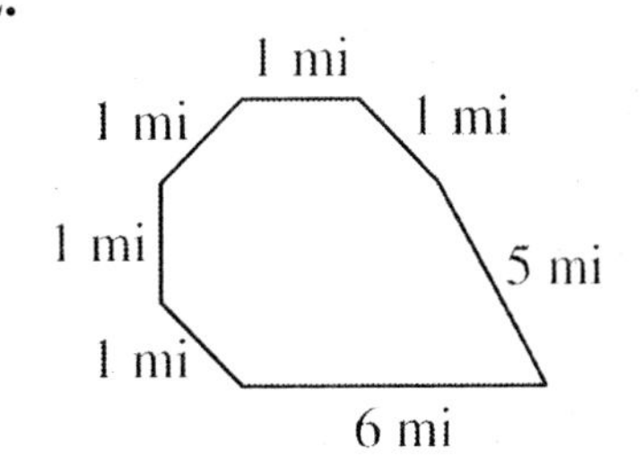

21. ____________

22. ____________

23. Radius = 4 in.

24. Radius = 45 cm

23. ________________

24. ________________

25. Radius = 10 in.

25. ________________

26. The figure below represents a walking path at a park. If Evan can complete the path in 19 minutes, what is Evan's average speed in meters per minute?

26. ________________

481 m
509 m
400 m
449 m
99 m

27. The figure in Exercise #17 represents a track at a community center. If Jean wants to walk 9600 m, how many times around the track does Jean have to walk?

27. ________________

28. Jean has a stride of 65 cm. How many steps does Jean take to walk 5200 m?

28. ________________

Concept Connections

29. The figure in Exercise #5 roughly represents a trail around a pond. If Sue walks at a rate of 50 meters per minute, how long would it take her to complete the trail?

____

____

30. The figure in Exercise #8 represents the path for a bike trail. If Anna can complete the course at an average speed of 16 miles per hour, how long did it take her to complete the course?

____

____

Name: Date:
Instructor: Section:

Chapter 7 PROBLEM SOLVING WITH GEOMETRY

Activity 7.4

Learning Objectives

1. Write formulas for areas of squares, rectangles, parallelograms, triangles, trapezoids, and polygons.
2. Calculate areas of polygons using appropriate formulas.

Key Terms

Use the vocabulary terms listed below to complete each statement in Exercises 1–5.

parallelogram **rectangle** **square** **trapezoid** **triangle**

1. The general formula for the area of a ____________________ is $A = s^2$.

2. The general formula for the area of a ____________________ is $A = \frac{1}{2}bh$.

3. The general formula for the area of a ____________________ is $A = \frac{1}{2}h(b+B)$.

4. The general formula for the area of a ____________________ is $A = bh$.

5. The general formula for the area of a ____________________ is $A = lw$.

Practice Exercises

For #6-17, find the area of each of the following figures.

6.

30 ft

80 ft

7.

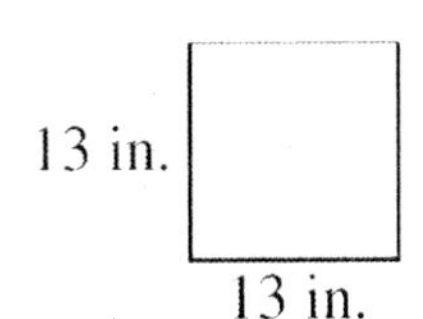

6. ________________

7. ________________

8.

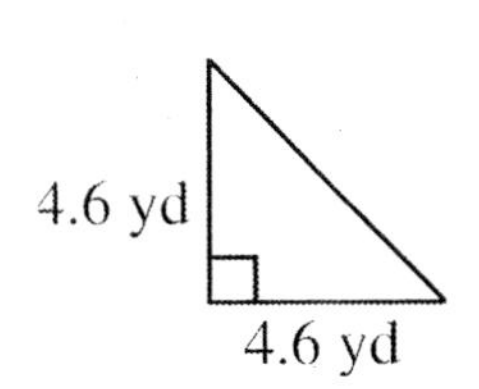

9.

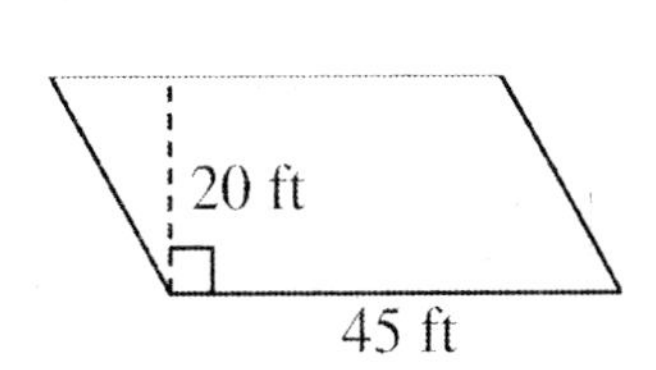

8. ________________

9. ________________

10.

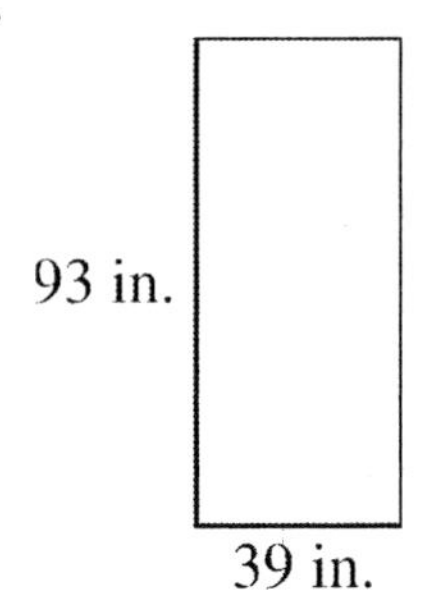

11.

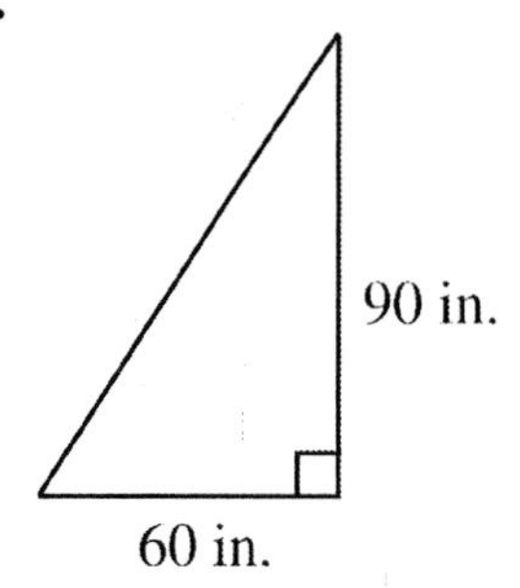

12.

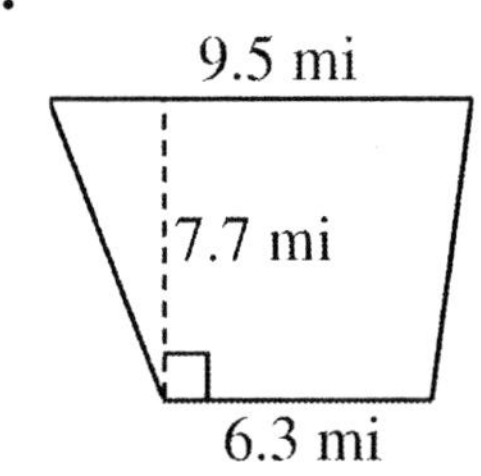

13.

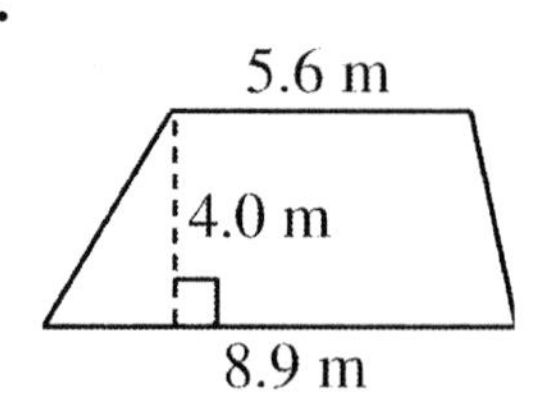

14.

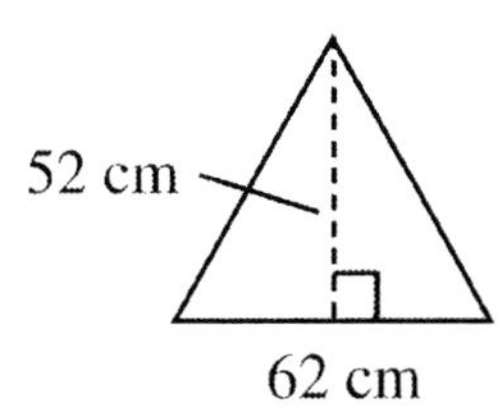

15.

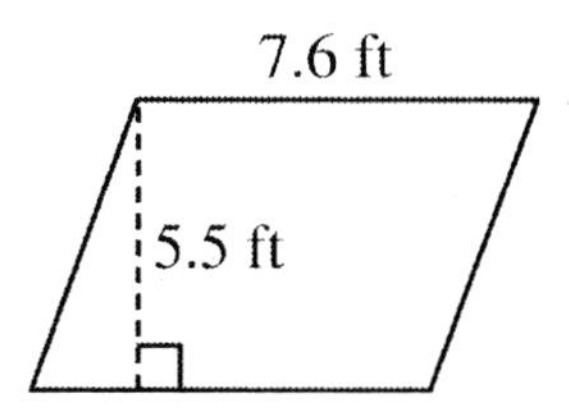

16.

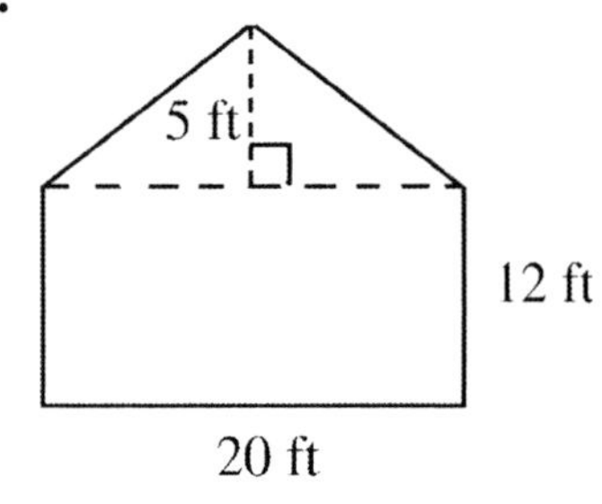

17.

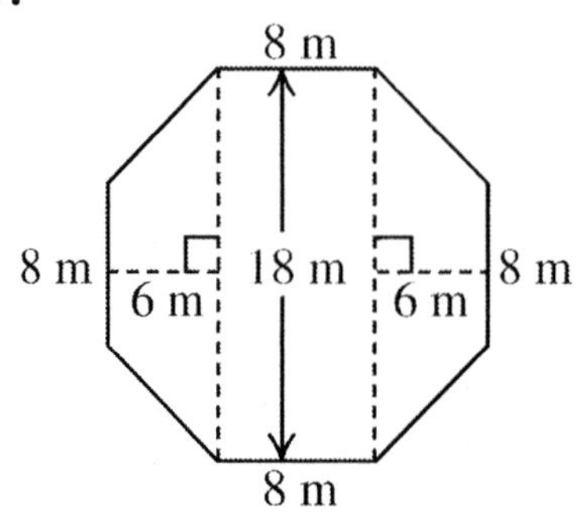

10. ________________

11. ________________

12. ________________

13. ________________

14. ________________

15. ________________

16. ________________

17. ________________

Name: Date:
Instructor: Section:

18. Find the area of a trapezoid that has height 6 mm and bases 14 mm and 18 mm.

18. ______________

19. Find the area of a triangle that has height 5 inches and base 12 inches.

19. ______________

20. A swimming pool is 50 m by 25 m. Find the area of the surface of the pool.

20. ______________

21. An Olympic size swimming pool is 164 ft by 82 ft. Find the area of the surface of the pool.

21. ______________

22. A football field is 360 feet by 160 feet. Find the area of the football field.

22. ______________

23. Your front lawn is a rectangle that is 27 feet wide and 40 feet long. Find the area of the lawn.

23. ______________

24. Your driveway is a rectangle that is 14 feet wide and 30 feet long. Find the area of the driveway.

24. ______________

25. Every year you cover your driveway (from Exercise #24) with sealant. If sealant costs $0.04 per square foot, how much would it cost to cover the driveway in sealant?

25. ______________

26. You plant a triangular flower garden that has base 8 feet and height 4.5 feet. Find the area of the garden.

26. 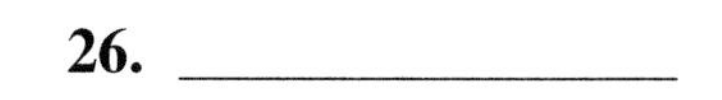

27. Every year you plant new annuals in a square garden with a 1.5-meter side. What is the area of the garden?

27. 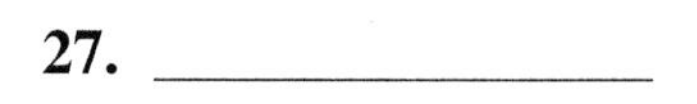

Concept Connections

This is the figure from Exercise #13. Suppose you want to buy carpet for this area that costs $2.50 a square meter and only comes in rectangular shaped pieces.

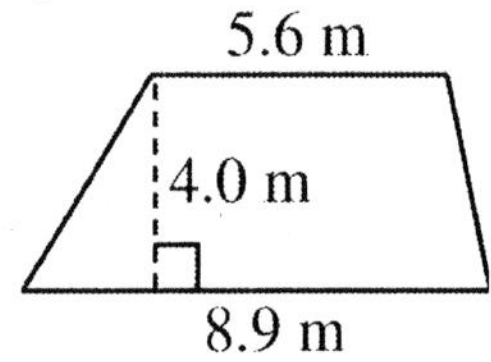

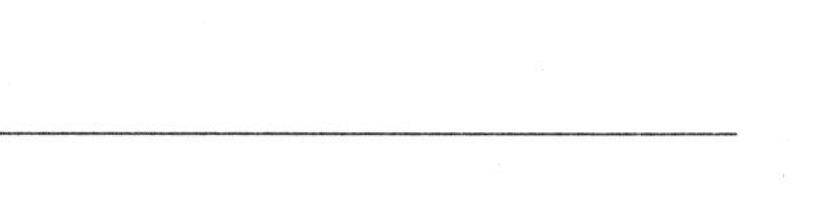

28. What are the dimensions of the carpet you would have to buy?

29. What would be the area of the piece of carpet you would have to buy? How much would it cost?

30. After putting down the carpet, it must be trimmed to create the correct trapezoidal shape. What is the area of the carpet that would be trimmed off and discarded?

Name: Date:
Instructor: Section:

Chapter 7 PROBLEM SOLVING WITH GEOMETRY

Activity 7.5

Learning Objectives
1. Develop formulas for the area of a circle.
2. Use the formulas to determine areas of circles.

Practice Exercises

For #1-9, find the area for each circle, given the radius. Round to the nearest hundredth.

1. radius = 12.5 in.

2. radius = 8.3 miles

3. radius = 30 cm

4. radius = 4.61 m

5. radius = 9 ft

6. radius = 7.9 mm

7. radius = 3 in.

8. radius = 5.5 ft

9. radius = 20 mi

1. ______________

2. ______________

3. ______________

4. ______________

5. ______________

6. ______________

7. ______________

8. ______________

9. ______________

For #10-17, find the area for each circle, given the diameter. Round to the nearest hundredth.

10. diameter = 63.2 m

11. diameter = 18 ft

12. diameter = 40.8 mm

13. diameter = 29 in.

14. diameter = 0.74 mi

15. diameter = 30 cm

16. diameter = 10 mi

17. diameter = 6 in.

10. ______________

11. ______________

12. ______________

13. ______________

14. ______________

15. ______________

16. ______________

17. ______________

For #18-23, find the area of each circle. Round to the nearest hundredth.

18.

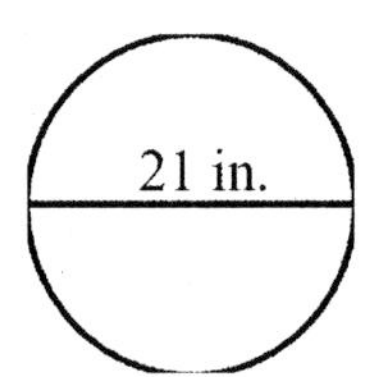

19.

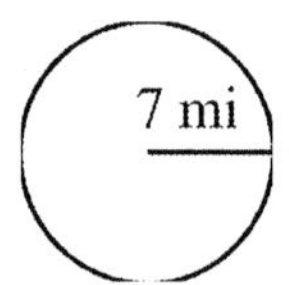

18. ______________

19. ______________

Name: Date:
Instructor: Section:

20.

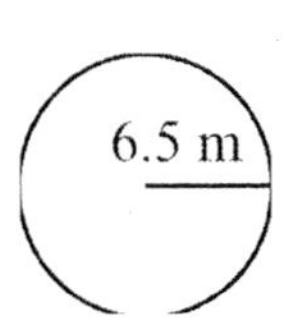

21.

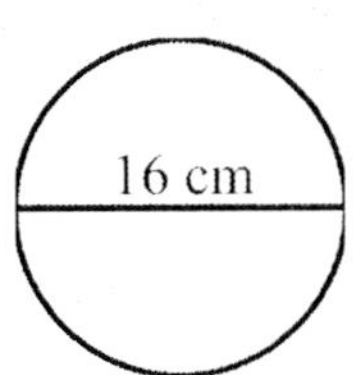

20. ______________

21. ______________

22.

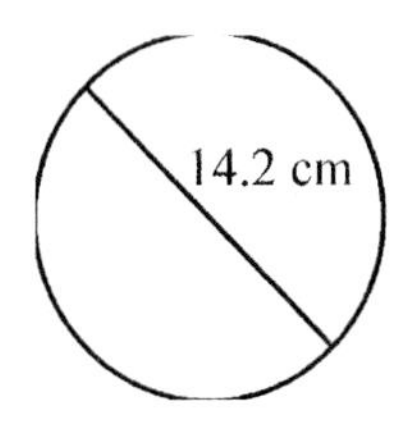

23.

22. ______________

23. ______________

24. A circle has area 84 sq. in. Find the radius. Round to the nearest tenth.

24. ______________

25. A circle has area 300 sq. ft. Find the radius. Round to the nearest tenth.

25. ______________

26. A circle has area 453 sq. mi. Find the radius. Round to the nearest mile.

26. ______________

27. A circle has area 1257 sq. cm. Find the radius. Round to the nearest cm.

27. ________________

28. A circle has area 14 sq. m. Find the radius. Round to the nearest tenth.

28. ________________

Concept Connections

29. An 8-inch apple pie contains 6 servings. A 10-inch pumpkin pie contains 8 servings. What is the area of one serving from each of the pies? Round to the nearest tenth. Which pie has the larger serving size, based on area?

____

____

30. For the area of a circle using the diameter, why does the formula include division by 4?

____

____

Name: Date:
Instructor: Section:

Chapter 7 PROBLEM SOLVING WITH GEOMETRY

Activity 7.6

Learning Objectives
1. Solve problems in context using geometric formulas.
2. Distinguish between problems that require area formulas and those that require perimeter formulas.

Key Terms
Use the vocabulary terms listed below to complete each statement in Exercises 1–2.

area **perimeter**

1. ____________________ formulas are used to measure the length around a plane figure.

2. ____________________ formulas are used to measure the amount of space inside a plane figure.

Practice Exercises

3. You want to build a fence around a play area in your yard that is 30 ft by 20 ft. What is the perimeter of this area? **3.** ________________

4. If fencing costs $12.50 per foot, how much would it cost to put a fence around a square-shaped area that is 225 square feet? **4.** ________________

5. To save money, you decide to buy cheaper fencing at $10.50 per foot, and to put the fence next to the house, so that only three sides of the 20 ft by 20 ft section require fencing. How much would this fencing cost? **5.** ________________

6. You want to paint two walls with an accent color. One wall is 15 ft by 11 ft. The other wall is 9 ft by 11 ft. What is the total area of the walls? **6.** ________________

7. A rectangular picture frame is 12 inches by 14 inches. What is the perimeter of the frame? **7.** ________________

8. A rectangular quilt is 90 inches by 100 inches. What is the perimeter of the quilt?

8. ______________

9. A rectangular quilt is 90 inches by 100 inches. What is the area of the quilt?

9. ______________

10. You buy material for the backing for a rectangular quilt that is 68 inches by 86 inches. The material only comes in whole square-yard units. How many square yards must you purchase? (144 square inches = 1 square foot and 9 square feet = 1 square yard)

10. ______________

11. A hallway is 28 ft by 4.5 ft. What is the area of the hallway?

11. ______________

12. Mid-grade carpeting is priced at $8 a square foot. How much would it cost to carpet a rectangular room that is 8.5 feet wide and 14 feet long?

12. ______________

13. Cheap carpeting is priced at $3.50 a square foot. How much would it cost to carpet a rectangular room that is 8.5 feet wide and 14 feet long?

13. ______________

14. It costs $11 per foot to have crown moulding professionally installed. How much would it cost to put in crown moulding for a room that is 17 feet by 30 feet?

14. ______________

15. It costs $11 per foot to have crown moulding professionally installed. How much would it cost to put in crown moulding for a room that is 12 feet by 10 feet?

15. ______________

16. You and your neighbor want to stain your decks. Your deck is 15 feet by 10 feet. Your neighbor's deck is 14 feet by 12 feet. What is the total area of the decks?

16. ______________

Name: Date:
Instructor: Section:

17. Stain is sold in 1 gallon containers that cover 250 square feet. How many gallons would you need to cover the total area of decks that are of size 20 feet by 20 feet, 10 feet by 15 feet, and 8 feet by 10 feet?

17. ____________

18. A microwaveable pizza sits on a disk that has a has a 6-inch diameter. Find the area of the disk.

18. ____________

19. The square box that holds the microwave pizza has a side of length 7 inches. What is the area of the bottom of the box?

19. ____________

20. A coffee can has a diameter of 4 inches. What is the area of the lid? Round to the nearest tenth.

20. ____________

21. A table top has a 40-inch diameter. What is the circumference? Round to the nearest tenth.

21. ____________

22. A table top has a 40-inch diameter. What is the area? Round to the nearest tenth.

22. ____________

23. The top of a rectangular toaster oven has length 16 inches and width 11 inches. What is the area?

23. ____________

24. The top of a rectangular toaster oven has length 16 inches and width 11 inches. What is the perimeter?

24. ____________

25. The lid of a briefcase is 12 inches by 16 inches. What is the perimeter?

25. ____________

26. The lid of a briefcase is 12 inches by 16 inches. What is the area?

26. ________________

27. A piece of notebook paper is 8.5 inches by 11 inches. What is the perimeter?

27. ________________

28. A piece of notebook paper is 8.5 inches by 11 inches. What is the area?

28. ________________

Concept Connections

29. What is the difference between the circumference and perimeter of a circle?

____

____

30. The length of a side of a square is 2. The area of the square is 4. The perimeter of the square is 4. Even though the values for area and perimeter the same, what kind of measurement do each represent?

____

____

Name: Date:
Instructor: Section:

Chapter 7 PROBLEM SOLVING WITH GEOMETRY

Activity 7.7

Learning Objectives

1. Verify and use the Pythagorean Theorem for right triangles.
2. Calculate the square root of numbers other than perfect squares.
3. Use the Pythagorean Theorem to solve problems.
4. Determine the distance between two points using the distance formula.

Practice Exercises

For #1-6, determine whether the three numbers represent a Pythagorean triple.

1. 21, 28, 35 **2.** 15, 36, 39

1. ________________

2. ________________

3. 8, 16, 17 **4.** 2, 6, 9

3. ________________

4. ________________

5. 7, 24, 25 **6.** 5, 11, 13

5. ________________

6. ________________

For #7-12, determine the distance between the given points.

7. (7, –1) and (4, 3) **8.** (2, 4) and (5, 8)

7. ________________

8. ________________

9. $(-10, -3)$ and $(14, 4)$

10. $(-4, 2)$ and $(2, 2)$

9. ______________

10. ______________

11. $(9, -5)$ and $(9, 2)$

12. $(-3, 2)$ and $(9, -3)$

11. ______________

12. ______________

For #13-20, determine the length of the missing side of each right triangle. Round to the nearest tenth.

13. leg = 18 cm,
hypotenuse = 21 cm

14. leg = 9 ft,
leg = 3 ft

13. ______________

14. ______________

15. leg = 5 mi,
leg = 10 mi

16. leg = 48 in.,
hypotenuse = 50 in.

15. ______________

16. ______________

17. leg = 2 m,
hypotenuse = 5 m

18. leg = 1 mm,
leg = 3 mm

17. ______________

18. ______________

19. leg = 7 yd,
leg = 8 yd

20. leg = 8 in.,
hypotenuse = 10 in.

19. ______________

20. ______________

Name: Date:
Instructor: Section:

For #21-18, determine the distance between the given points. Round to the nearest tenth.

21. (3, 5) and (2, –2)

22. (–2, –5) and (4, 7)

23. (2, 2) and (–4, –1)

24. (–2, –1) and (4, 3)

25. (–1, 3) and (3, –5)

26. (5, 1) and (–1, 2)

27. (–1, 1) and (5, 3)

28. (–2, 7) and (1, 4)

21. ______________

22. ______________

23. ______________

24. ______________

25. ______________

26. ______________

27. ______________

28. ______________

Concept Connections

29. Draw a right triangle. Label the right angle, legs and hypotenuse.

30. State the Pythagorean Theorem.

____

____

Name: Date:
Instructor: Section:

Chapter 7 PROBLEM SOLVING WITH GEOMETRY

Activity 7.8

Learning Objectives

1. Recognize geometric properties of three-dimensional figures.
2. Write formulas for and calculate surface areas of rectangular prisms (boxes), right circular cylinders (cans), and spheres (balls).

Practice Exercises

For #1-11, find the surface area of each of the following figures. Round to the nearest square unit.

1.

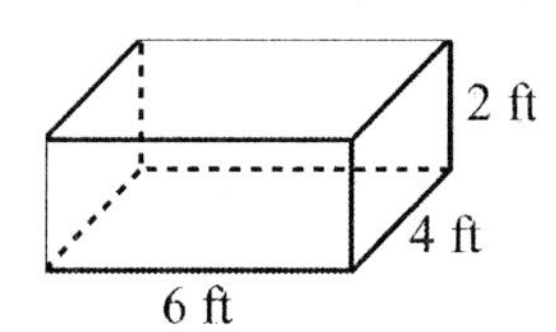

2.

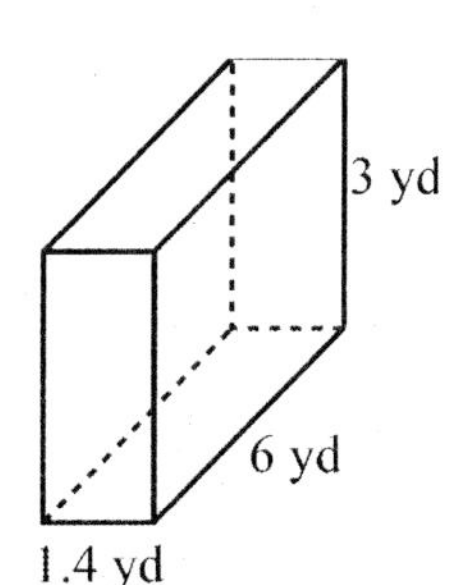

1. ________

2. ________

3.

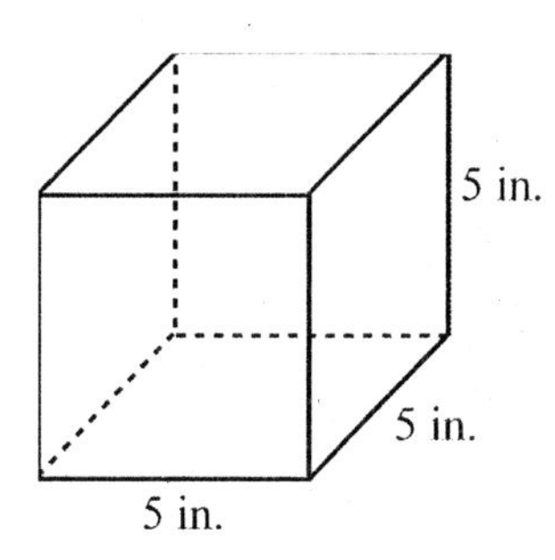

4.

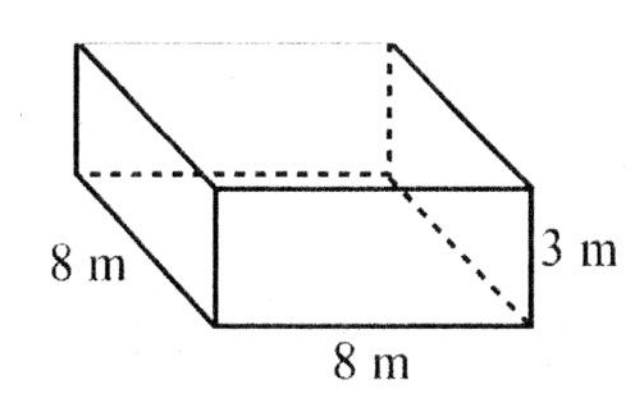

3. ________

4. ________

5.

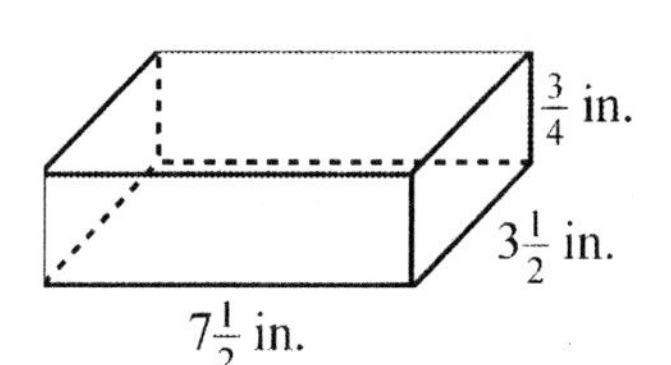

6.

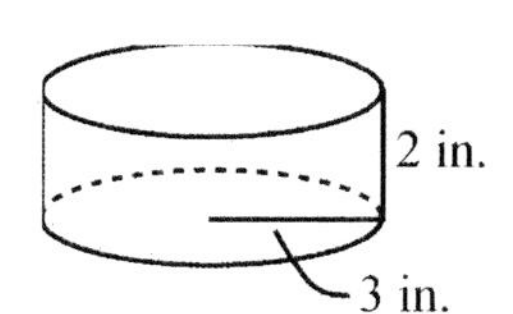

5. ________

6. ________

7.

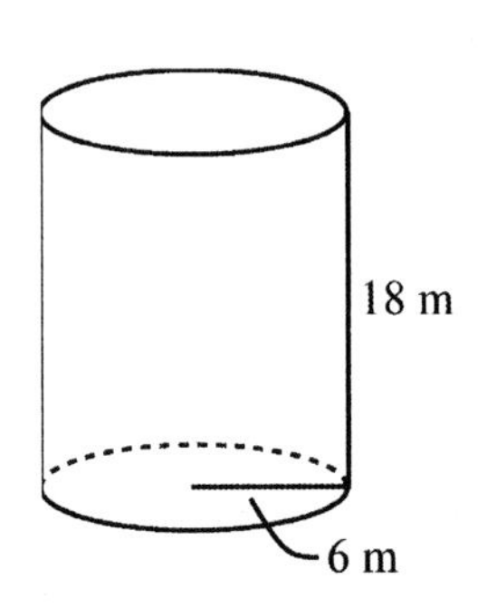

8.

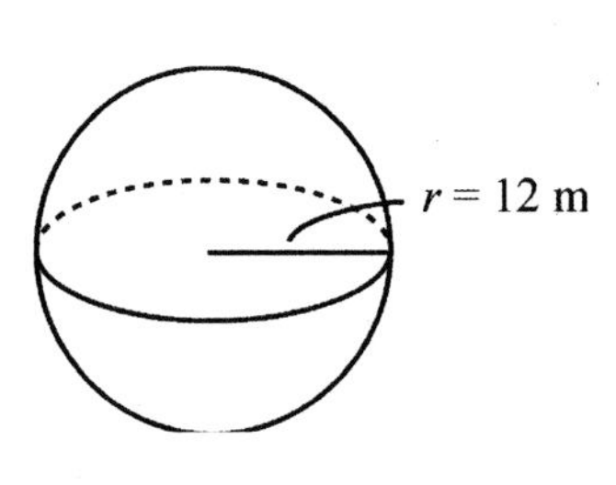

7. ________________

8. ________________

9.

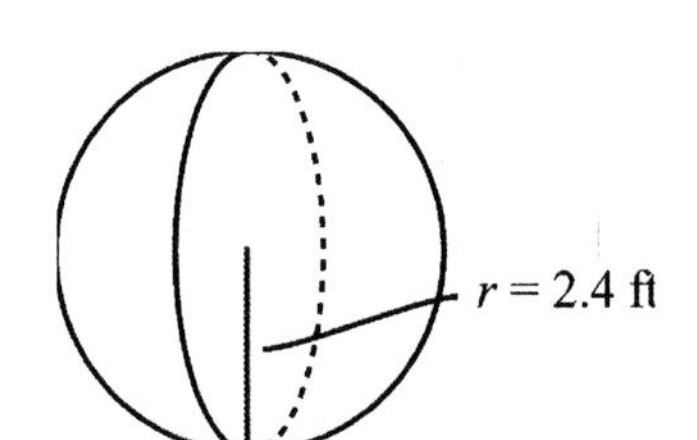

10.

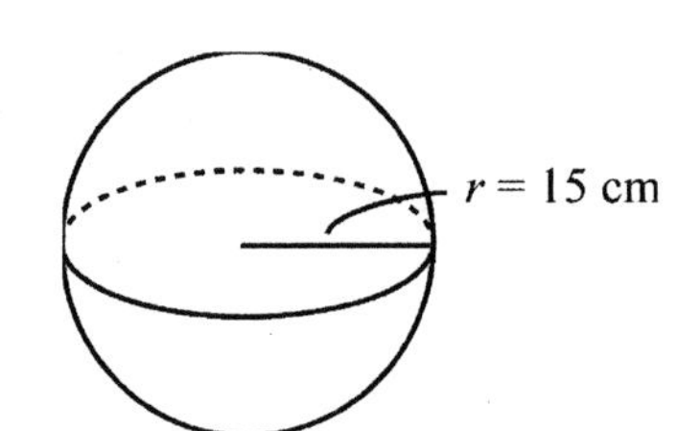

9. ________________

10. ________________

11.

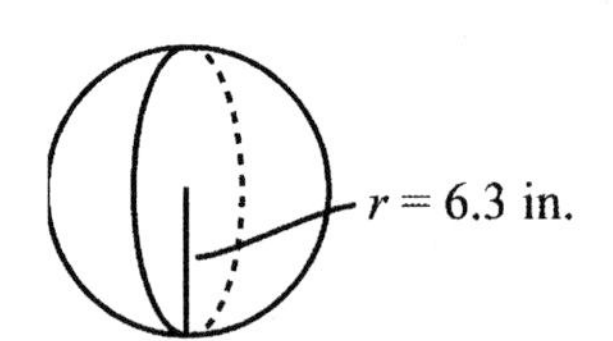

11. ________________

For #12-17, find the surface area of the right circular cylinder with the given dimensions. Round to the nearest square unit.

12. Radius is 1 foot and height is 2 feet

13. Radius is 10 m and height is 40 m

12. ________________

13. ________________

Name: Date:
Instructor: Section:

14. Radius is 5 inches and height is 9.5 inches

15. Radius is 14 cm and height is 25 cm

14. ______________

15. ______________

16. Radius is 36 inches and height is 2 inches

17. Radius is 5 mm and height is 20 mm

16. ______________

17. ______________

For #18-23, find the surface area of the rectangular prism with the given dimensions. Round to the nearest square unit.

18. Length is 5 inches, width is 5 inches, and height is 3 inches

19. Length is 10.5 yards, width is 3.5 yards, and height is 1 yard

18. ______________

19. ______________

20. Length is 2 feet, width is 1.5 feet, and height is 0.75 feet

21. Length is 40 cm, width is 35 cm, and height is 20 cm

20. ______________

21. ______________

22. Length is 12 inches, width is 8 inches, and height is 8 inches

23. Length is 12 feet, width is 10 feet, and height is 10 feet

22. ______________

23. ______________

For #24-26, find the surface area of the sphere with the given dimension. Round to the nearest square unit.

24. Radius is 3 inches

25. Radius is 5.5 feet

24. ______________

25. ______________

26. Radius is 20 mm

26. ______________

27. The surface area for a new roll of paper towels is 510.5 square inches. If the radius of the roll is 5 inches, how tall is the roll? Round to the nearest hundredth.

27. ______________

28. The surface area for a jar of peanut butter is 208.9 square inches. If the radius of the jar is 3.5 inches, how tall is the jar?

28. ______________

Concept Connections

29. A box without a lid has height 6 inches, width is 8 inches, and length 14 inches. What is the surface area of the box?

____

____

30. You are to wrap a bowling ball that has radius 4.5 inches. What is the least amount of wrapping paper you must buy in order to complete the job?

____

____

Name: Date:
Instructor: Section:

Chapter 7 PROBLEM SOLVING WITH GEOMETRY

Activity 7.9

Learning Objectives

1. Write formulas for and calculate volumes of rectangular prisms (boxes) and right circular cylinders (cans).
2. Recognize geometric properties of three-dimensional figures.

Key Terms

Use the vocabulary terms listed below to complete each statement in Exercises 1–4.

volume **cubic** **rectangular prism** **right circular cylinder**

1. The formula for the space enclosed by a ________________ is given by $V = \pi r^2 h$.

2. The ________________ is the space inside a three-dimensional figure.

3. The formula for the space enclosed by a ________________ is given by $V = lwh$.

4. The space inside a three-dimensional figure is measured in ________________ units.

Practice Exercises

For #5-12, find the volume of the right circular cylinder with the given dimensions. Round to the nearest cubic unit.

5. Radius is 1 foot and height is 2 feet

6. Radius is 10 m and height is 40 m

7. Radius is 5 inches and height is 9.5 inches

8. Radius is 14 cm and height is 25 cm

9. Radius is 36 inches and height is 2 inches

10. Radius is 5 mm and height is 20 mm

5. ________________

6. ________________

7. ________________

8. ________________

9. ________________

10. ________________

11. Radius is 2 yards and height is 1 yard

12. Radius is 4 miles and height is 1.5 miles

11. ______________

12. ______________

For #13-19, find the volume of the rectangular prism with the given dimensions. Round to the nearest cubic unit.

13. Length is 5 inches, width is 5 inches, and height is 3 inches

14. Length is 10.5 yards, width is 3.5 yards, and height is 1 yard

13. ______________

14. ______________

15. Length is 2 feet, width is 1.5 feet, and height is 0.75 feet

16. Length is 40 cm, width is 35 cm, and height is 20 cm

15. ______________

16. ______________

17. Length is 12 inches, width is 8 inches, and height is 8 inches

18. Length is 12 feet, width is 10 feet, and height is 10 feet

17. ______________

18. ______________

19. Length is 80 mm, width is 10 mm, and height is 30 mm

19. ______________

Name: Date:
Instructor: Section:

For #20-26, find the volume of each figure. Round to the nearest cubic unit.

20.

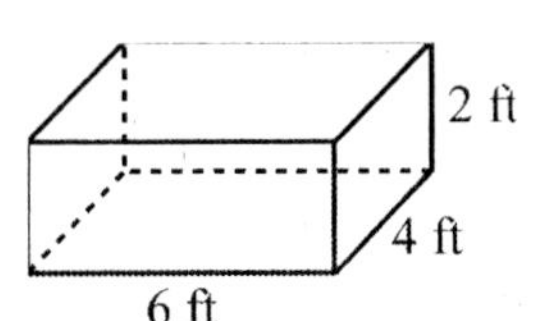

21.

3 yd
6 yd
1.4 yd

20. 

21. ____________

22.

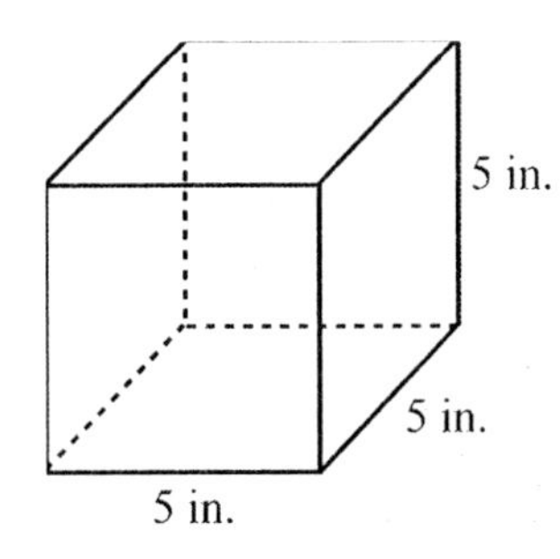

23.

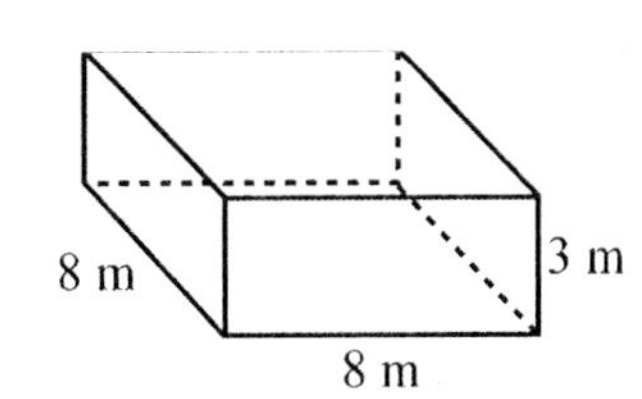

22. ____________

23. ____________

24.

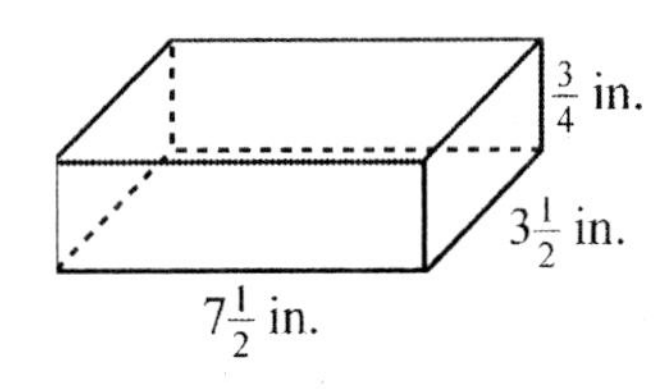

25.

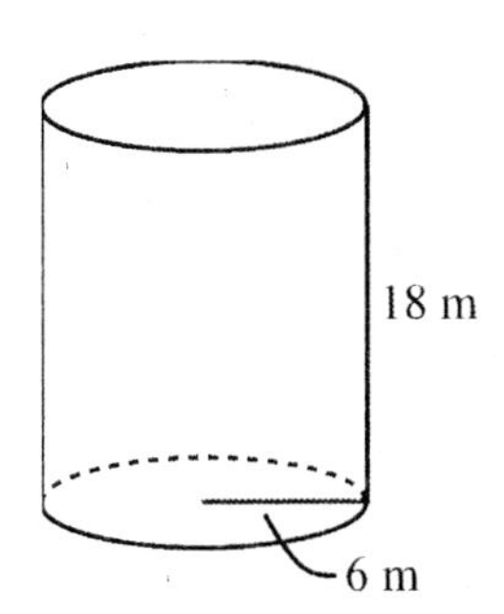

24. ____________

25. ____________

26.

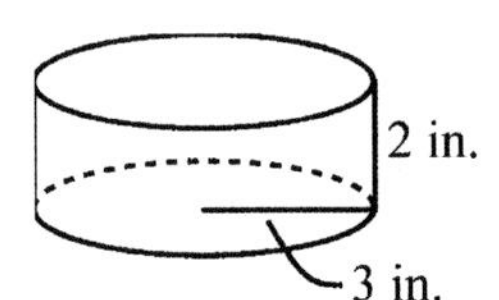

26. 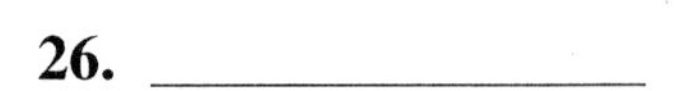

27. The volume for a can of coffee is 163.6 cubic inches. If the radius of the can is 3.5 inches. How tall is the can? Round to the nearest hundredth.

27. ________________

28. The volume for a can of cranberry sauce is 95 cubic inches. If the radius of the can is 2.75 inches. How tall is the can? Round to the nearest inch.

28. ________________

Concept Connections

29. Two different glasses are used in a fruit and vegetable-themed restaurant to serve beverages. The Celery Glass is 8 inches tall and has a 3-inch diameter. The Pineapple Glass is 4 inches tall and has a 4-inch diameter. What are the volumes of each of the glasses? Which glass has the larger volume?

____

____

30. A toy cube is to be wrapped in padding an inch thick around the toy. If each side of the toy is 4 inches long, what is the volume of the toy covered in padding?

____

____

Name: Date:
Instructor: Section:

Chapter 8 PROBLEM SOLVING WITH MATHEMATICAL MODELS

Activity 8.1

Learning Objectives

1. Describe a mathematical situation as a set of verbal statements.
2. Translate verbal rules into symbolic equations.
3. Solve problems that involve equations of the form $y = ax + b$.
4. Solve equations of the form $y = ax + b$ for the input x.
5. Evaluate the expressions $ax + b$ in an equation of the form $y = ax + b$ to obtain the output y.

Practice Exercises

For #1-10, solve each of the following equations for x.

1. $17 = 2x + 5$

2. $-37 = -4x - 17$

3. $4x - 13 = 15$

4. $26 - 3x = 38$

5. $6x - 13 = 26$

6. $-3x + 7 = 7$

7. $0.45x - 1.6 = 2$

8. $6 = 3.6x - 30$

1. ________________

2. ________________

3. ________________

4. ________________

5. ________________

6. ________________

7. ________________

8. ________________

9. $13+\frac{1}{4}x=8$

10. $\frac{3}{4}x-15=0$

9. ________________

10. ________________

For #11-20, replace y by the given value and solve the resulting equation for x.

11. If $y = 4x - 6$ and $y = 18$, determine x.

12. If $y = 20 - 3x$ and $y = 11$, determine x.

11. ________________

12. ________________

13. If $y=\frac{2}{3}x-23$ and $y=-17$, determine x.

14. If $y=20-\frac{5}{4}x$ and $y=5$, determine x.

13. ________________

14. ________________

15. If $2.4x-7=y$ and $y=5$, determine x.

16. If $-4x+25=y$ and $y=-3$, determine x.

15. ________________

16. ________________

Name: Date:
Instructor: Section:

17. If $y = 7 - 11x$ and $y = -4$, determine x.

18. If $6x - 17 = y$ and $y = 10$, determine x.

17. ______________

18. ______________

19. If $y = 2.5x - 2$ and $y = 8$, determine x.

20. If $y = -3x + 15$ and $y = 0$, determine x.

19. ______________

20. ______________

For #21-28, complete the table.

21. $y = 5x - 9$

x	y
7	
	56

22. $y = 15 + 1.5x$

x	y
0.5	
	-15

23. $y = -4x + 25$

x	y
$\frac{3}{4}$	
	-7

24. $y = 13 - \frac{4}{5}x$

x	y
-10	
	1

25. $y = \frac{x}{6} - 3$

x	y
36	
	12

26. $y = 8 - 2.4x$

x	y
5	
	−13

27. $y = 5x - 9$

x	y
2	
	−29

28. $y = -2x - 7$

x	y
−8	
	−7

Concept Connections

29. Explain in your own words how to solve $y = 3x - 5$ for y, given $x = 2$. Then give the answer.

____

____

30. Explain in your own words how to solve $y = 3.2x + 2$ for x, given $y = 10$. Then give the answer.

____

____

Name: Date:
Instructor: Section:

Chapter 8 PROBLEM SOLVING WITH MATHEMATICAL MODELS

Activity 8.2

Learning Objectives

1. Write symbolic equations from information organized in a table.
2. Produce tables and graphs to compare outputs from two different mathematical models.
3. Solve equations of the form $ax + b = cx + d$.

Practice Exercises

For #1-28, solve each equation.

1. $7x-6=4x+12$

2. $5x-11=7x+23$

1. ______________

2. ______________

3. $0.4x+8=5.4x-12$

4. $5x-12=-3x+4$

3. ______________

4. ______________

5. $0.4x-6.5=0.3x+3.2$

6. $5-0.03x=0.2-0.05x$

5. ______________

6. ______________

7. $8t+6=5t+15$

8. $5w-15=w-11$

7. ______________

8. ______________

9. $3+4x=5x+30$

10. $-81+x=7-15x$

9. ________________

10. ________________

11. $11+15x=7x-13$

12. $-2+18x=17x-1$

11. ________________

12. ________________

13. $43-7x=12-6x$

14. $3x+6=8x+11$

13. ________________

14. ________________

15. $-57=-7x+20$

16. $93=-x+4$

15. ________________

16. ________________

17. $0.4x+3=5.4x+18$

18. $1.4x-15=6-1.6x$

17. ________________

18. ________________

Name: Date:
Instructor: Section:

19. $0.7x - 3.3 = 0.6x + 11.4$

20. $5 - 0.015x = 0.035x$

21. $10 - 0.06x = 1.9 - 0.05x$

22. $5x - 9 + 3 = 7x - 11$

23. $4x - 9.6 = 7.4$

24. $71 - x = 14 - 6x$

25. $2100 - 4.7x = 20 + 5.3x$

26. $32 + 4x = 10.5 + 3x$

27. $2x + 5 = 5x - 2$

28. $8 - 3x = 8x - 3$

19. ______________

20. ______________

21. ______________

22. ______________

23. ______________

24. ______________

25. ______________

26. ______________

27. ______________

28. ______________

Concept Connections

29. A mobile phone company offers two monthly plans. Plan A includes 1000 texts and unlimited calls for $54.99 plus $0.10 for every additional text in excess of 1000. Plan B includes unlimited texts and unlimited calls for $69.99. Determine the number of texts for which the cost would be the same from both companies.
Hint: write an equation for the cost of each plan, and then write as a single equation.

30. You rent a car for the day. Company A charges $25.99 a day and $0.64 for each mile driven. Company B charges $55.44 a day and $0.33 for each mile driven. Determine the mileage for which the cost would be the same from both companies.

Name: Date:
Instructor: Section:

Chapter 8 PROBLEM SOLVING WITH MATHEMATICAL MODELS

Activity 8.3

Learning Objectives
1. Develop an equation to model and solve a problem.
2. Solve problems using formulas as models.
3. Recognize patterns and trends between two variables using a table as a model.
4. Recognize patterns and trends between two variables using a graph as a model.

Practice Exercises

For #1-4, the formula used to determine the cost of a cab ride, in dollars, is $C(m) = 1.5m + 2$*, where m is in miles.*

1. Find the cost of a cab ride that is 5 miles.

2. You have $13.25. How far will a cab take you?

3. Find the cost of a cab ride that is 3 miles.

4. You have $8.00. How far will a cab take you?

1. ______________

2. ______________

3. ______________

4. ______________

For #5-8, the formula used to determine the annual salary in dollars, of a book sales representative is $S(b) = 0.95b + 20{,}000$ *where b is the number of books sold.*

5. Find the annual salary of a book sales representative who sells 6000 books.

6. If the annual salary of the book sales representative was $44,700, how many books were sold?

7. Find the annual salary of a book sales representative who sells 8000 books.

8. If the annual salary of the book sales representative was $34,250, how many books were sold?

5. ______________

6. ______________

7. ______________

8. ______________

For #9-24, solve.

9. If $y = x + 9$ and $x = -17$, find y.

10. If $y = x + 9$ and $y = 14$, find x.

9. ______________

10. ______________

11. If $y = x + 7.3$ and $x = -3.9$, find y.

12. If $y = x + 7.3$ and $y = 19.7$, find x.

11. ______________

12. ______________

13. If $z = -13.5x$ and $x = -24$, find z.

14. If $z = -13.5x$ and $z = 332.1$, find x.

13. ______________

14. ______________

15. If $y = x + 17$ and $x = -21$, find y.

16. If $y = x + 17$ and $y = -24$, find x.

15. ______________

16. ______________

17. If $z = x - 21$ and $x = -7$, find z.

18. If $z = x - 21$ and $z = -11$, find x.

17. ______________

18. ______________

Name: Date:
Instructor: Section:

19. If $y = x + 2.3$ and $x = -0.7$, find y.

20. If $y = x + 2.3$ and $y = 3.8$, find x.

19. ______________

20. ______________

21. If $y = 14.6x$ and $x = 12$, find y.

22. If $y = 14.6x$ and $y = -306.6$, find x.

21. ______________

22. ______________

23. If $P = 2l + 2w$ and $l = 3,\ w = 5$, find P.

24. If $P = 2l + 2w$ and $l = 65,\ P = 150$, find w.

23. ______________

24. ______________

For #25-27, use the following information. Grapes are $1.50 per pound, including tax.

25. How much would you pay for 4 pounds of grapes?

26. Use appropriate letters to represent the variables involved and translate the written statement to an equation.

25. ______________

26. ______________

27. Complete the table.

Pounds of grapes	2	3	6	8	10
Cost					

28. A brand new car is valued at $24,000. Each year the value of the car decreases by $2000. Complete the table.

Year	2	4	6	8	10
Value of car					

Concept Connections

29. Using the information from #28, is zero a reasonable replacement value for input?

____

____

30. What is the purpose of mathematical models?

____

____

Odd Answers

Chapter 1 WHOLE NUMBERS

Activity 1.1

Key Terms

1. Rounding
3. Even
5. factors
7. place value

Practice Exercises

9. 6,291,834
11. 852,023
13. 563,300
15. 10,000,000
17. >
19. <
21. two hundred seven thousand nine hundred thirty-four
23. 23,663 ends with an odd digit
25. 95 is composite since 95 has factors other than 1 and 95, namely 5 and 19.
27. 23 is prime since 23 has no factors other than 1 and 23.

Concept Connections

29. The common difference in consecutive even whole numbers is 2.
    The common difference in consecutive odd whole numbers is also 2.

Activity 1.2

Key Terms

1. vertical; output
3. horizontal; input

Practice Exercises

5. 20
7. 120, 140, 160 and 180
9. (12, 20)
11. 16

13–19.

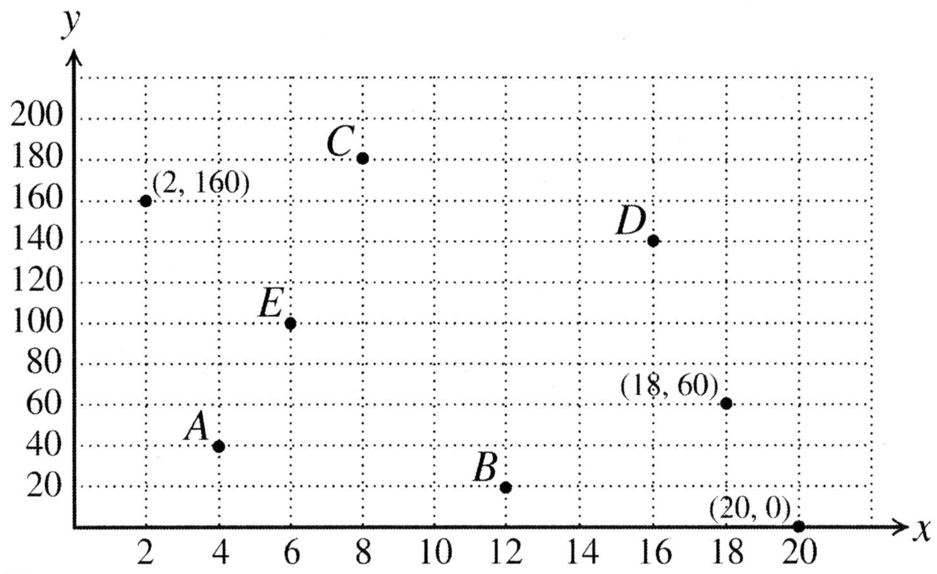

20–23.

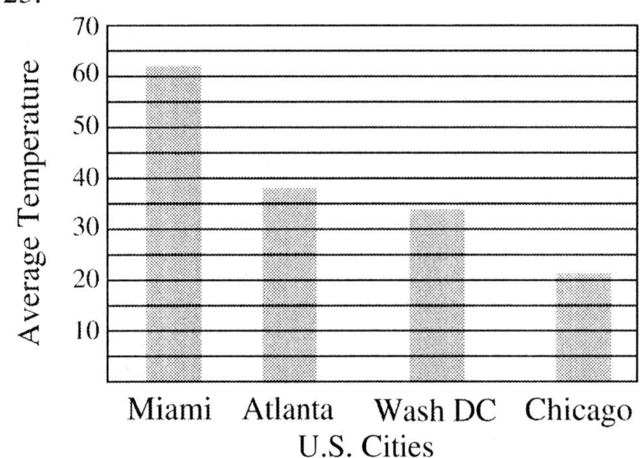

25. Quit smoking, 5%

27. Being more active, 55%

Concept Connections

29. In an ordered pair, the order is important. The first value is the x-coordinate and the second value is the y-coordinate.

Activity 1.3

Key Terms

1. subtrahend
3. addends
5. associative
7. commutative

Practice Exercises

9. 963
11. 981
13. 122
15. Yes
17. 116
19. Yes

21. 600
23. 326
25. 16 + 37
27. 250 – 63

Concept Connections

29. The estimate was slightly higher than the actual sum.
    Yes, it is possible. Find the differences between the rounded values and the exact values.

Activity 1.4

Key Terms

1. product; factor
3. distributive

Practice Exercises

5. 144
7. 534
9. 2795
11. 2720
13. 342
15. 672
17. 2000 napkins
19. 500 toothpicks
21. Commutative property
23. 4080
25. 2646
27. Yes

Concept Connections

29. The first grader used addition. The fourth-grader used multiplication and the commutative property of multiplication. The eight-grader used the distributive property.

Activity 1.5

Key Terms

1. quotient; divisor; remainder; dividend
3. zero
5. one

Practice Exercises

7. quotient: 39; remainder: 0
9. quotient: 20; remainder 3
11. quotient: 25; remainder: 22
13. undefined
15. quotient: 33; remainder: 6
17. 9 packets each and 3 left over
19. 9 subtractions of 5 with 3 left over
21. No
23. 125
25. higher
27. quotient: 55; remainder 285

Concept Connections

29. Purchase 5 packages of hotdogs and 7 packages of rolls. There will be 6 rolls leftover.

Activity 1.6

Key Terms

1. square

3. exponent; base; exponential form; power
5. perfect square
7. one

Practice Exercises

9. 100,000,000
11. 10^{10}
13. 71, 73, 79
15. $4 \cdot 1 = 2 \cdot 2 = 4$
17. $51 \cdot 1 = 17 \cdot 3 = 51$
19. 5 different factorizations
21. $2^2 \cdot 5$
23. $2^2 \cdot 3^3$
25. 49
27. 6^{11}

Concept Connections

29. The Fundamental property of whole numbers: For any whole number, there is only one prime factorization.

Activity 1.7

Practice Exercises

1. 204
3. 204
5. Exercises #1 and #3
7. 24
9. 21
11. 9
13. 10
15. 5
17. 125
19. 7
21. 36
23. 23
25. 54
27. 5

Concept Connections

29. The Commutative Property does not hold for subtraction or division.

Chapter 2 VARIABLES AND PROBLEM SOLVING

Activity 2.1

Key Terms

1. area
3. formula
5. constant

Practice Exercises

7. 256 sq. in.
9. 513 sq. ft.
11. 323 sq. cm
13. 86 cm
15. 1008 sq. ft.
17. 2 cans
19. 77°F
21. 10°C
23. $y = x + 57$
25. $y = 8x$
27. $y = 4(x - 11)$

Concept Connections

29. The perimeter is the distance around a figure. The area is the measure of the size of the region bounded by the sides of the figure.

Activity 2.2

Key Terms

1. input; output
3. ordered pair
5. graph

Practice Exercises

7. The x variable is the input.

9.

C degrees Celsius	F degrees Fahrenheit
0	32
5	41
10	50
15	59
20	68

11. The letter S

13. The letter x

15. The letter s

17.

t	0	1	2	3	4	5	6	7
h	0	96	160	192	192	160	96	0

19. I think it will reach 200 ft.

21. At 7 seconds

23.

t	0	1	2	3	4	5
d	0	65	130	195	260	325

25. 3 hours

27. 4 hours

Concept Connections

29. When $t = 9$, then $h = -288$. Nine seconds after the launch, the rocket is 288 ft below ground. This is not possible.

Activity 2.3

Key Terms

1. fundamental principle of equality
3. solved

Practice Exercises

5. $x = 134$
7. $w = 170$
9. $t = 0$
11. $x = 87$
13. $w = 2869$
15. $x = 520$
17. $y = 0$
19. $x = 727$
21. $s = 207$
23. $x = 20$

25. $m = 0$

27. $y = 362$

Concept Connections

29. An expression can be evaluated and simplified and does not have an equals sign. An equation includes an equals sign and can be solved.

Activity 2.4

Practice Exercises

1. $x = 9$
3. $y = 50$
5. $z = 14$
7. $w = 2$
9. $t = 0$
11. $x = 37$
13. $x = 22$
15. $w = 239$
17. $s = 15$
19. $g = 3$
21. $r = 38$
23. $z = 1385$
25. $t = 5429$
27. $w = 12$

Concept Connections

29. If $a = 0$ in the equation $ax = b$, after substitution, the equation would be $0 \cdot x = b$, which simplifies to $0 = b$. Then we cannot solve for x.

Activity 2.5

Practice Exercises

1. 3 terms
3. $6x$ and $7x$
5. 4 terms
7. $8a$ and $8a$; $6b$ and $8b$
9. $11s + 26t + 25$
11. $14b + 14c - 17$
13. $26w + 185z + 78y$
15. $y = 20$
17. $w = 0$
19. $t = 2$
21. $x = 26$
23. $x = 39$
25. $y = 7$
27. $x = 28$

Concept Connections

29. To check your answer, substitute the value for the variable into the original equation and simplify. If both sides of the equation are the same, the answer is correct.

Activity 2.6

Practice Exercises

1. $x + 361 = 504$; $x = 143$
3. $x - 627 = 132$; $x = 759$
5. $x - 60 = 32$; $x = 92$
7. $19 + x = 43$; $x = 24$
9. $791 = 7x$; $x = 113$
11. $x + 669 = 883$; $x = 214$

13. $228 = 38x$; $x = 6$
15. $x + 532 = 3607$; $x = 3075$
17. $53x = 689$; $x = 13$
19. $x - 51 = 23$; $x = 74$
21. $833 = 7x$; $x = 119$
23. $97 + x = 112$; $x = 15$
25. I must drive 65 mph.
27. I must save $225 per month.

Concept Connections

29. Step 1. Understand the Problem. Step 2. Develop a strategy for solving the problem. Step 3. Execute your strategy to solve the problem. Step 4. Check your solution for correctness.

Chapter 3 PROBLEM SOLVING WITH INTEGERS

Activity 3.1

Key Terms

1. negative; positive

Practice Exercises

3. $>$
5. $<$
7. $>$
9. 6
11. 59
13. 100
15. 11
17. –3
19. –43
21. –55 points
23. –10 yards
25. –22°
27. $130

Concept Connections

29. On the number line, the absolute value of a number is represented by the distance the number is from zero.

Activity 3.2

Practice Exercises

1. –15
3. –12
5. –7
7. –19
9. –18
11. 9
13. 0
15. –11
17. –5
19. –3
21. –12
23. –21
25. 0
27. 8

Concept Connections

29. The difference is 14,776 ft.

Activity 3.3

Practice Exercises

1. $A = B - 314$
3. $R = P + C$
5. $A = B + C$
7. $A = C - 54$
9. $C = F - D$
11. $x + 12 = -15$; $x = -27$
13. $x - 10 = 25$; $x = 35$
15. $x - 29 = -6$; $x = 23$
17. $x - 35 = 5$; $x = 40$
19. $3 - x = 9$; $x = -6$
21. $c = 44$
23. $a = 135$
25. $c = 56$
27. $a = 11$

Concept Connections

29. $x = y + 2b - x$ is not solved for x since there is an x on both sides of the equation.

Activity 3.4

Key Terms

1. quadrants
3. origin

Practice Exercises

5. second quadrant
7. fourth quadrant
9. y-axis
11. x-axis
13. second quadrant
15. fourth quadrant
17. third quadrant
19. x-axis
21. (2, 5)
23. (0, –1)
25. $x - y = 8$
27. $x + y = -2$

Concept Connections

29. On the first graph, the points form a straight line. On the second graph they form a zig zag pattern, depending on the spacing between tick marks.

Activity 3.5

Practice Exercises

1. 48
3. –21
5. 0
7. 28
9. 5
11. not possible
13. –6
15. 72
17. 110
19. not possible
21. 64
23. 1
25. 169
27. 30

Concept Connections

29. $-3 + 2 + (-5) + 1 + (-8) = -13$. The change was a decrease of 13 ft. The snow pack level after week 5 was 23 ft.

Activity 3.6

Practice Exercises

1. 7
3. -80
5. -25
7. -12
9. -8
11. 7
13. -112
15. 69
17. $x = 7$
19. $y = 3$
21. $8a + 12b$
23. $6x + 2y$
25. $x = 8$
27. $y = -11$

Concept Connections

29. The change in height is $\frac{-6\text{ feet}}{6\text{ days}}$, or -1 foot per day.

Chapter 4 PROBLEM SOLVING WITH FRACTIONS

Activity 4.1

Key Terms

1. rational number
3. lowest terms
5. greatest common factor

Practice Exercises

7. 20
9. 63
11. 24
13. 9
15. $\frac{1}{5}$
17. $\frac{4}{13}$
19. $>$
21. $>$
23. $<$
25. 30
27. 72

Concept Connections

29. The denominator b, cannot be zero because division by zero is undefined.

Activity 4.2

Key Terms

1. one
3. multiplying

5. dividing

Practice Exercises

7. $\frac{21}{40}$

9. $\frac{1}{32}$

11. $\frac{1}{4}$

13. 2

15. -11

17. $-\frac{1}{30}$

19. $-\frac{1}{6}$

21. $x=-\frac{14}{15}$

23. $y=-\frac{5}{16}$

25. $t=6$

27. $\frac{3}{4}x=-6;\ x=-8$

Concept Connections

29. To divide one fraction by another fraction, multiply the dividend fraction by the reciprocal of the divisor fraction.

Activity 4.3

Practice Exercises

1. $\frac{7}{10}$

3. $\frac{2}{3}$

5. 0

7. $\frac{2}{5}$

9. $-\frac{7}{8}$

11. $\frac{3}{8}$

13. $-\frac{1}{6}$

15. $-\frac{2}{45}$

17. $\frac{1}{24}$

19. $c=\frac{1}{2}$

21. $x=\frac{14}{15}$

23. $x=-\frac{13}{15}$

25. $b=-\frac{11}{16}$

27. $w=-\frac{1}{5}$

Concept Connections

29. Identify the largest denominator of the fractions involved. Then look at multiples of the largest denominator. The smallest multiple that is divisible by each of the denominator(s) is the LCD.

Activity 4.4

Practice Exercises

1. $\frac{9}{25}$
3. $\frac{25}{81}$
5. $\frac{9}{10}$
7. 5
9. undefined
11. 29
13. $\frac{8}{27}$
15. $x = \frac{8}{35}$
17. $t = \frac{2}{3}$
19. $y = \frac{5}{8}$
21. $x = -\frac{3}{4}$
23. $x = 10$
25. $b = -\frac{2}{3}$
27. $t = 1$ second

Concept Connections

29. If $b = 0$, then there would be division by zero, which is undefined.

Chapter 5 PROBLEM SOLVING WITH MIXED NUMBERS AND DECIMALS

Activity 5.1

Key Terms

1. improper fraction

Practice Exercises

3. $1\frac{1}{5}$
5. $-2\frac{1}{4}$
7. $-1\frac{1}{3}$
9. $-5\frac{1}{4}$
11. $\frac{47}{8}$
13. $-\frac{27}{4}$
15. $-\frac{14}{3}$
17. $\frac{31}{4}$
19. $1\frac{3}{7}$
21. $1\frac{1}{6}$
23. $2\frac{1}{4}$
25. $-2\frac{1}{13}$

27. $-7\frac{1}{3}$

Concept Connections

29. $-5\frac{8}{9}=-5+\left(-\frac{8}{9}\right)$

Activity 5.2

Practice Exercises

1. $14\frac{3}{10}$
3. $3\frac{1}{8}$
5. $20\frac{7}{36}$
7. $13\frac{11}{12}$
9. $8\frac{1}{5}$
11. $22\frac{8}{9}$
13. $1\frac{21}{26}$
15. $-14\frac{2}{15}$
17. $4\frac{2}{3}$
19. $x=12\frac{1}{6}$
21. $x=4\frac{3}{5}$
23. $x=3\frac{5}{12}$
25. $x=-12\frac{5}{12}$
27. $x=6\frac{7}{8}$

Concept Connections

29. Alice did not find the least common denominator before adding the numerators.

    The solution is $2\frac{2}{3}+3\frac{7}{9}=2+\frac{6}{9}+3+\frac{7}{9}=5\frac{13}{9}=6\frac{4}{9}$.

Activity 5.3

Practice Exercises

1. $25\frac{1}{5}$
3. $-27\frac{17}{39}$
5. 16
7. $2\frac{7}{10}$
9. $1\frac{1}{7}$
11. $-2\frac{1}{2}$
13. $\frac{16}{25}$
15. $\frac{64}{81}$

17. $\frac{5}{9}$

19. $\frac{13}{6}$ or $2\frac{1}{6}$

21. $9\frac{1}{3}$

23. $\frac{3}{4}$

25. $x = 10$

27. $x = -\frac{1}{8}$

Concept Connections

29. Tony can see a maximum of 28 km.

Activity 5.4

Practice Exercises

1. thirty-five thousandths
3. five hundred sixty and one hundred thirty-two ten-thousandths
5. 0.0073
7. 643.00911
9. 0.1
11. 0.70
13. >
15. >
17. 0.85
19. 0.25
21. 0.467
23. 0.714
25. 0.818
27. 2.5

Concept Connections

29. Irrational numbers include nonterminating and nonrepeating decimals.

Activity 5.5

Practice Exercises

1. 18.68
3. –0.064
5. 2.329
7. –21.718
9. 16.195
11. –0.00182
13. –0.231
15. –1.267
17. $x = 5.364$
19. $x = 100.48$
21. $w = -5.86$
23. $b = 0$
25. $x = 390.33$
27. $x = 58.933$

Concept Connections

29. The difference is 101°F.

Activity 5.6

Practice Exercises

1. 87.504
3. –16.731

5. –3.0096
7. 1.938
9. –0.39325
11. 2.9
13. 56.350
15. 0.36
17. –116,413.5
19. 4.72
21. 150
23. $150,000
25. $9000
27. 4500

Concept Connections

29. Rounding 390.56 to 400 and 4.05 to 4, the estimated product is 1600. No, his answer is not correct. The correct answer is 1581.7680.

Activity 5.7

Practice Exercises

1. 51.67
3. 19.9464
5. 1.689991
7. 0.278
9. 189.63
11. 7.5
13. $\frac{x}{7.2} = -5.1$; $x = -36.72$
15. $23.8 = x \cdot (-3.5)$; $x = -6.8$
17. $(-8.3)x = 49.8$; $x = -6$
19. $x = 3.5$
21. $x = 1.2$
23. $x = 23.6$
25. $x \approx 1.2$
27. $x \approx 1.7$

Concept Connections

29. First replace each variable on the right-hand side with its given value.
 $3357 = P + P(0.085)(1.4)$
 Then solve the equation for P using the order of operations. $P = \$3000$.

Activity 5.8

Key Terms

1. centigram
3. milligram

Practice Exercises

5. 9.62 m
7. 7.629 m
9. 8730 m
11. 0.843 kg
13. 0.982 L
15. 6.752 m
17. 6.572 km
19. 8310 m
21. 950 g
23. 19 mL
25. 9170 g
27. 6,950,000 mm

Concept Connections

29. The basic metric unit for distance is meter. The basic metric unit for mass is gram. The basic metric unit for volume is liter.

Chapter 6 PROBLEM SOLVING WITH RATIOS, PROPORTIONS, AND DECIMALS

Activity 6.1

Key Terms
1. ratio
3. percent
5. verbal

Practice Exercises
7. $\frac{55}{103}$
9. $\frac{13}{20}$
11. $\frac{99}{100}$
13. 0.85
15. 0.25
17. 0.467
19. 74%
21. 43.75%
23. 12.5%
25. 0.045
27. 0.0063

Concept Connections
29. The missing information is the number of baskets attempted for each player.

Activity 6.2

Practice Exercises
1. $x = 24$
3. $x = 30$
5. $x = 54$
7. $x = 522$
9. $x = 1600$
11. $x = 195$
13. 374 miles
15. 84,000 miles
17. 875 shares
19. 105 games
21. False
23. True
25. False
27. False

Concept Connections
29. Bill is right. After cross multiplication, both proportions are equivalent to $a \cdot d = b \cdot c$.

Activity 6.3

Key Terms
1. total
3. proportional reasoning

Practice Exercises
5. 32
7. 24
9. 81
11. 8

13. 800
15. 40
17. 324
19. 19,000
21. 1400
23. 96.6
25. 17,500
27. 3048

Concept Connections

29. There are 408 right-handed children.

Activity 6.4

Key Terms

1. relative change

Practice Exercises

3. $15
5. $24
7. $20
9. history
11. 3.2%
13. 39.1%
15. 35 lbs
17. 35 lbs
19. 6
21. 13
23. lack of water
25. 6.5%
27. 6.1%

Concept Connections

29. The percent decrease is 75%. No, the amount of decrease cannot exceed the original amount.

Activity 6.5

Key Terms

1. multiplying
3. growth factor

Practice Exercises

5. 140% = 1.40
7. 103% = 1.03
9. 322% = 3.22
11. 100.2% = 1.002
13. 100.5% = 1.005
15. 193% = 1.93
17. 1.0875
19. $1.75
21. 1.041
23. 24,598
25. 215%
27. $55,200

Concept Connections

29. The new value is larger since the growth factor is greater than 1.

Activity 6.6

Key Terms

1. dividing
3. decay factor, subtracting

Practice Exercises

5. 60% = 0.60
7. 97% = 0.97
9. 78% = 0.78
11. 1% = 0.01
13. 99.5% = 0.995
15. 7% = 0.07
17. 0.946
19. 16.7%
21. $119.99
23. 200 pounds
25. 0.95
27. 0.75

Concept Connections

29. The original value is larger since the decay factor is less than 1.

Activity 6.7

Key Terms

1. product

Practice Exercises

3. 0.75
5. $95.63
7. 0.90
9. $56,782.69
11. 20 lbs
13. $2772
15. 0.275
17. 0.21
19. $373,763
21. $84,248
23. 0.75
25. Decay factor, since the value is less than 1.
27. 6.25% decrease

Concept Connections

29. The cumulative effect for the first deal is 49% decrease. The cumulative effect for the second deal is 44% decrease. The first deal is better.

Activity 6.8

Practice Exercises

1. 54,000 seconds
3. 2.485 miles
5. 70.87 inches
7. 25,200 minutes
9. 1.44 grams/30 days
11. 4.5 yards
13. 190,080 inches
15. 0.42 grams/12 weeks
17. 138,336 feet
19. 23.1 lb
21. 10.6 kpL
23. 2.11 qt
25. 600 ft
27. 504 hours

Concept Connections

29. All the French recipes used Metric system, and Julia had to convert everything to American units for her cookbook. Since she didn't use conversion rate analysis, she had to replicate each recipe through trial and error, which was very time consuming.

Chapter 7 GEOMETRY

Activity 7.1

Key Terms

1. trapezoid
3. rectangle

Practice Exercises

5. 220 ft
7. rectangle
9. 52 in.
11. square
13. 15.7 yd
15. isosceles triangle
17. 31.6 mi
19. 80 cm
21. 264 in.
23. 168 cm
25. 24 ft
27. 20 ft

Concept Connections

29. A parallelogram is a closed four-sided plane figure whose opposite sides are parallel and the same length. A trapezoid is a closed four-sided plane figure with two opposite sides that are parallel and two opposite sides that are not parallel.

Activity 7.2

Key Terms

1. radius
3. diameter

Practice Exercises

5. 16.6 miles
7. 9.22 m
9. 15.8 mm
11. 9 ft
13. 14.5 in.
15. 15 cm
17. 128.2 mm
19. 56.5 ft
21. 198.5 m
23. 44.0 mi
25. 50.3 cm
27. 88.0 ft

Concept Connections

29. The perimeter of a circle is the same as its circumference. $C = \pi d$ or $C = 2\pi r$

Activity 7.3

Practice Exercises

1. 45 cm
3. 23.2 in.
5. 150 m
7. 69.1 ft
9. 95 yd
11. 22.3 in.
13. 122 ft
15. 108 cm
17. 1600.0 m
19. 154 in.
21. 1931 m
23. 26.8 in.
25. 51.4 in.
27. 6 times around

Concept Connections

29. It would take Sue 3 minutes to complete the trail.

Activity 7.4

Key Terms

1. square
3. trapezoid
5. rectangle

Practice Exercises

7. 169 sq. in.
9. 900 sq. ft.
11. 2700 sq. in.
13. 29 sq. m
15. 41.8 sq. ft.
17. 300 sq. m
19. 30 sq. in.
21. 13,448 sq. ft.
23. 1080 sq. ft.
25. $16.80
27. 2.25 sq. m

Concept Connections

29. The area of the carpet would be 35.6 square meters and would cost $89.

Activity 7.5

Practice Exercises

1. 490.87 sq. in.
3. 2827.43 sq. cm
5. 254.47 sq. ft
7. 28.27 sq. in.
9. 1256.64 sq. mi.
11. 254.47 sq ft
13. 660.52 sq. in.
15. 706.86 sq. cm
17. 28.27 sq. in.
19. 153.94 sq. mi
21. 201.06 sq. cm
23. 615.75 sq. ft
25. 9.8 ft
27. 20 cm

Concept Connections

29. For the apple pie, one serving has area 8.4 sq. in. For the pumpkin pie, one serving has area 9.8 sq. in. The pumpkin pie has the larger serving size.

Activity 7.6

Key Terms

1. perimeter

Practice Exercises

3. 100 ft.
5. $630
7. 52 in.
9. 9000 sq. in.
11. 126 sq. ft.
13. $416.50
15. $484
17. 3 gallons
19. 49 sq. in.
21. 125.7 in
23. 176 sq. in.
25. 56 in.
27. 39 in.

Concept Connections

29. There is no difference. The perimeter is the same as the circumference of a circle.

Activity 7.7

Practice Exercises

1. Yes
3. No
5. Yes
7. 5
9. 25
11. 7
13. leg = $\sqrt{117} \approx 10.8$ cm
15. hypotenuse = $\sqrt{125} \approx 11.2$ mi
17. leg = $\sqrt{21} \approx 4.6$ m
19. hypotenuse = $\sqrt{113} \approx 10.6$ yd
21. $\sqrt{50} \approx 7.1$
23. $\sqrt{45} \approx 6.7$
25. $\sqrt{80} \approx 8.9$
27. $\sqrt{40} \approx 6.3$

Concept Connections

29. The figure below is a right triangle.

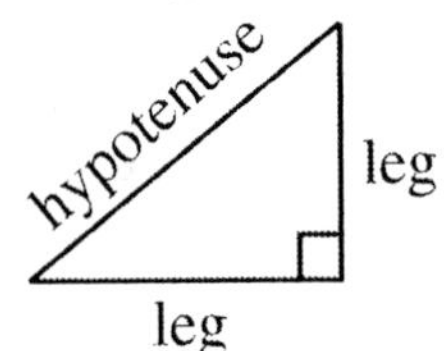

Activity 7.8

Practice Exercises

1. 88 square feet
3. 150 square inches
5. 69 square inches
7. 905 square meters
9. 72 square feet
11. 499 square inches
13. 3142 square meters
15. 3431 square centimeters
17. 785 square millimeters
19. 102 square yards

21. 5800 square centimeters
23. 680 square feet
25. 380 square feet
27. 11.25 inches

Concept Connections

29. The surface area is 376 square inches.

Activity 7.9

Key Terms

1. right circular cylinder
3. rectangular prism

Practice Exercises

5. 6 cubic feet
7. 746 cubic inches
9. 8143 cubic inches
11. 13 cubic yards
13. 75 cubic inches
15. 2 cubic feet
17. 768 cubic inches
19. 24,000 cubic millimeters
21. 25 cubic yards
23. 192 cubic meters
25. 2036 cubic meters
27. 4.25 inches

Concept Connections

29. The Celery Glass is 56.5 cu. in. The Pineapple Glass is 50.3 cu. in. The Celery Glass has the larger volume.

Chapter 8 PROBLEM SOLVING WITH MATHEMATICAL MODELS

Activity 8.1

Practice Exercises

1. $x = 6$
3. $x = 7$
5. $x = 6.5$
7. $x = 8$
9. $x = -20$
11. $x = 6$
13. $x = 9$
15. $x = 5$
17. $x = 1$
19. $x = 4$

21.

x	y
7	26
13	56

23.

x	y
$\frac{3}{4}$	22
8	−7

25.

x	y
36	3
90	12

27.

x	y
2	1
−4	−29

Concept Connections

29. Substitute 2 for x in the equation and evaluate. The answer is $y = 1$.

Activity 8.2

Practice Exercises

1. $x = 6$
3. $x = 4$
5. $x = 97$
7. $t = 3$
9. $x = -27$
11. $x = -3$
13. $x = 31$
15. $x = 11$
17. $x = -3$
19. $x = 147$
21. $x = 810$
23. $x = 4.25$
25. $x = 208$
27. $x = \frac{7}{3}$

Concept Connections

29. 1150 texts

Activity 8.3

Practice Exercises

1. \$9.50
3. \$6.50
5. \$25,700
7. \$27,600
9. $y = -8$
11. $y = 3.4$
13. $z = 324$
15. $y = -4$
17. $z = -28$
19. $y = 1.6$
21. $y = 175.2$
23. $P = 16$
25. \$6
27.

Pounds of grapes	2	3	6	8	10
Cost	\$3.00	\$4.50	\$9.00	\$12.00	\$15.00

Concept Connections

29. Yes, when the input value is zero, the car is brand new.